Prometheanism

Critical Perspectives on Theory, Culture and Politics

Critical Perspectives on Theory, Culture and Politics is a new interdisciplinary series developed in partnership with the Centre for Critical and Cultural Theory based in the School of English, Communication and Philosophy at Cardiff University, UK. This interdisciplinary series will focus on innovative research produced at the interface between critical theory and cultural studies. In recent years much work in Cultural Studies has increasingly moved away from directly critical-theoretical concerns. One of the aims of this series is to foster a renewed dialogue between Cultural Studies and Critical and Cultural Theory in its rich, multiple dimensions.

Series Editors

Glenn Jordan, Reader in Cultural Studies and Creative Practice and Director of Butetown History & Arts Centre, University of South Wales.

Laurent Milesi, Reader in English, Communication and Philosophy and Chair of the Centre for Critical and Cultural Theory, Cardiff University.

Radhika Mohanram, Professor of English and Critical and Cultural Theory, Cardiff University.

Chris Norris, Distinguished Research Professor, Cardiff University.

Chris Weedon, Chair of the Centre for Critical and Cultural Theory, Director of Postgraduate Studies and Head of the Centre for Critical and Cultural Theory, Cardiff University.

Culture Control Critique: Allegories of Reading the Present, Frida Beckman

Prometheanism: Technology, Digital Culture and Human Obsolescence, Christopher John Müller

Creole in the Archive: Imagery, Presence and the Location of the Caribbean Figure, Roshini Kempadoo (forthcoming)

The Attention Economy: Labour, Time, and Power in Cognitive Capitalism, Claudio Celis (forthcoming)

Postcolonial Nostalgia and the Construction of a South-Asian Diaspora, Anindya Raychaudhuri (forthcoming)

Cultures of the Extreme: From Abu Ghraib to 'Saw' and Beyond, Pramod K. Nayar (forthcoming)

Credo Credit Crisis: Speculations on Faith and Money, edited by Laurent Milesi, Christopher John Müller and Aidan Tynan (forthcoming)

Materialities of Sex in a Time of HIV: The Promise of Vaginal Microbicides, Annette-Carina van der Zaag (forthcoming)

Performative Contradiction and the Romanian Revolution, Jolan Bogdan (forthcoming)

Partitions and their Afterlives: Violence, Memories, Living, edited by Radhika Mohanram and Anindya Raychaudhuri (forthcoming)

Music, Photography, and the Aesthetics of Time, Peter R. Sedgwick and Kenneth Gloag (forthcoming)

Contested Borders: Queer Politics and Cultural Translation in Contemporary Francophone Writing from the Maghreb, William J. Spurlin (forthcoming)

Affective Connections: Towards a New Materialist Politics of Sympathy, Dorota Golańska (forthcoming)

Chinese Subjectivity and the Beijing Olympics, Gladys Pak Lei Chong (forthcoming)

Prometheanism

Technology, Digital Culture and Human Obsolescence

Christopher John Müller

ROWMAN &
LITTLEFIELD
INTERNATIONAL
London • New York

Published by Rowman & Littlefield International, Ltd.
Unit A, Whitacre Mews, 26-34 Stannary Street, London SE11 4AB
www.rowmaninternational.com

Rowman & Littlefield International, Ltd. is an affiliate of Rowman & Littlefield
4501 Forbes Boulevard, Suite 200, Lanham, Maryland 20706, USA
With additional offices in Boulder, New York, Toronto (Canada), and Plymouth (UK)
www.rowman.com

Originally published under the title 'Über prometheische Scham' in *Die Antiquiertheit des Menschen Bd. I* by Günther Anders.

British Library Cataloguing in Publication Data
A catalogue record for this book is available from the British Library

ISBN: HB 978-1-78348-238-2
PB 978-1-78348-239-9

Library of Congress Cataloging-in-Publication Data
Names: Müller, Christopher John, author.
Title: Prometheanism : technology, digital culture and human obsolescence / Christopher John Müller.
Description: Lanham : Rowman & Littlefield International, 2016. | Series: Critical perspectives on theory, culture and politics | Includes bibliographical references and index.
Identifiers: LCCN 2016027658 (print) | LCCN 2016028904 (ebook) | ISBN 9781783482382 (cloth : alk. paper) | ISBN 9781783482399 (pbk. : alk. paper) | ISBN 9781783482405 (electronic)
Subjects: LCSH: Anders, Gunther, 1902–1992. | Human beings. | Philosophical anthropology. | Technology—Philosophy. | Technology—Moral and ethical aspects.
Classification: LCC B3199.A534 M85 2016 (print) | LCC B3199.A534 (ebook) | DDC 193—dc23
LC record available at https://lccn.loc.gov/2016027658

∞™ The paper used in this publication meets the minimum requirements of American National Standard for Information Sciences—Permanence of Paper for Printed Library Materials, ANSI/NISO Z39.48-1992.

Printed in the United States of America

Contents

Acknowledgements

This book is testament to the tremendous encouragement of my parents Shirley and Urs Müller, who have always offered their support unreservedly and unconditionally. For this I am truly grateful. The School of English, Communication and Philosophy and The Centre for Critical and Cultural Theory at Cardiff University enabled me to conduct this work in a highly stimulating and collegial environment and provided me with scarce office space. I wish to thank all my friends and colleagues here at Cardiff, too many to list individually, for the many discussions that have fed into this project. I do, however, especially wish to thank Chris Weedon whose support and friendship have been instrumental to the completion of this project. Her diligent readership and criticism have shaped this work from the outset and have improved it incalculably. Christopher Norris's helpful comments on previous versions of parts of this book, and Marija Grech's critical readership have helped me structure and refine my argument. Special mention must go to Aidan Tynan, who drew my attention to the publication of 'The Pathology of Freedom' in *Deleuze Studies* in 2009, as this was my first encounter with Anders's thought. Beyond Cardiff, I wish to thank Rodney Livingstone for his guidance and encouragement towards taking the first steps as a translator, and also for taking the time to revise parts of my translation. The friendship and critical awareness of Julius Greve and Kostas Dolgeras have made a major contribution to the overall trajectory of this book, and I would also like to thank Johann Gregory, Stefan Karrer and Edgar Eckert in this respect. This project would not have been possible without the work on Anders in English that a number of pioneering individuals have produced. I especially wish to mention Harold Marcuse's website on Anders, and Jason Dawsey's introductions to Anders's life and work, which have proven to be indispensable resources. I would also like to thank Kerstin Putz and Reinhard Ellensohn, the organisers of the 2014 FWF conference on Anders '*Schreiben für übermorgen*', and the participants of the said conference. Work presented there has been further developed in this book and an earlier, short version of the final chapter will appear in German in the conference volume. Etienne Morel has provided invaluable assistance with the translation of the Latin and Greek expressions that form part of Anders's 'On Promethean Shame'. Lucy Menon and Shirley Müller's many editorial and stylistic suggestions have significantly enhanced the overall finish and readability of this book. Finally, I would like to offer my sincere thanks to Daniel Karrer who has allowed me to use one of his wonderful paintings as a book cover.

Introduction

Thinking Finitude, Digital Technology and Human Obsolescence with Günther Anders

> 'Technology' is a fetish-word that covers over our lack of understanding of finitude and our terror at the precipitate and unbridled character of our 'mastery', which no longer knows either end or completion.[1]
>
> **Jean-Luc Nancy**

> Prometheus's triumph has been all too overwhelming.[2]
>
> **Günther Anders**

Prometheus, 'the brilliant', 'the clever, shifty one' first enters the Western imaginary as a trickster. The name Prometheus means 'fore-thought' and records the Titan's scheming and cunning nature, whereas his dim-witted brother Epimetheus ('after-thought') is described less flatteringly by Hesiod as 'the mind that missed the mark'.[3] In ancient mythology the story of Prometheus has many versions. Prometheus appears as the creator of mankind who bestows on us his brilliance and his ability to think and plan ahead. Other versions, Plato's *Protagoras*, for instance, which is central to Bernard Stiegler's influential reading in *Technics and Time 1*, focus on Epimetheus, the idiot brother without whom the figure of Prometheus, as Stiegler puts it, '*makes no sense*'.[4] While helping his brother to create life, Epimetheus forgets to reserve any of the qualities and abilities he distributed so harmoniously among animals, leaving the newly formed humans 'naked, unshod, without bedding or weapons'.[5] A third trope of the myth is given central stage in Aeschylus's account, in which Prometheus is shown to have protected us from the knowledge of death that foresight inevitably brings. As the Titan proudly declares with bound hands and a liver awaiting the eagle's tormenting beak, 'I stopped mortals from brooding on death', 'I made to lodge within them blind hopes'.[6]

These divergent accounts all converge on the central deed that the myth narrates, the deed that names *the source of the blind hopes* Aeschylus evokes: Prometheus tricks the gods and steals fire from them to bestow it on us. The gift of fire signifies technology and the skill to create and inhabit artificial structures with which Prometheus retroactively furnishes his creation. It is only once we have received these gifts that we can fend for ourselves. Humanity begins by becoming technological, by advancing into the space of possibility that technology and artificial skill open. As Pierre Hadot has outlined in a compelling reading, the trickery of Prometheus is remembered in the Greek *mekhane*, from which we derive the words 'machine' and 'mechanics': 'For the Greeks, mechanics first appeared as a technique for tricking nature, and by obliging nature to do what it cannot do by itself, by means of artificial and fabricated instruments, or "machines" – scales, wedges, screws, gears – which can serve, for instance, for the construction of war machines or automata.'[7] Hadot's words accentuate how Prometheus's trickery escapes the grasp and control of the trickster. The art and skill of tricking nature is not only employed to prolong life, to overcome suffering with medicine and to seek meaning in art and philosophy. The innate limitations and the frailty of the human body are not only reconfigured to extend a helpful hand, the power of artifice also begins to trick the good intentions of benevolent Prometheus. For, technology also names the innumerable ways in which humans are tricked out of lives worth living. It kills, destroys and exploits; it is used to create automata that can *replace* the creatures Prometheus sought to provide for and protect. 'You are pleased at having stolen fire and outwitted me', Hesiod's Zeus addresses the fire-thief with thundering words, 'a great calamity both for yourself and for humans to come'.[8]

This, of course, is not a book about Greek myth. Yet at the threshold of what Erik Brynjolfsson and Andrew McAfee have recently called *The Second Machine Age*, an age in which 'smart devices', 'the big data revolution' and 'networked and artificial intelligence' are reconfiguring all aspects of the consumerist societies in which they proliferate, the trickery of Prometheus opens us to ways of thinking about technology that resist the intellectually comfortable position of mobilising *a false opposition* between 'humanity' and 'technology' when looking ahead into our digital future.[9] Placed in opposition to humanity, 'technology' turns into a 'fetish word', to use Jean-Luc Nancy's phrase. The word 'technology' often takes on the function of a blanket description for a wide range of instruments, technological devices and machines that are considered to be under human control. The slogan of the American National Rifle Association 'guns don't kill people – people kill people' succinctly introduces the idea about technology which, as Arthur Bradley puts it, has 'dominated' Western thought: technological objects are neutral because they can be 'utilised for good or ill depending upon who or what happens to wield [them]'.[10]

At the same time and at the other end of the spectrum, 'technology' is opposed to 'humanity' in order to name a malignant force running out of control and threatening everything human. Here, the *instrumental definition* of technology just described is inverted to present machines as being in control of humanity. Such notions of technology becoming increasingly hostile towards their onetime masters find expression in the dystopian futures with which film and literature present us, and in the cultural anxieties that surround the prospect of artificial intelligence (in all its guises). This hostile position is frequently also narrowly ascribed to philosophical approaches that confront us with the idea that human existence has become progressively alienated in the course of our technological modernity (reductive readings of Karl Marx, Walter Benjamin and Martin Heidegger spring to mind). Our 'incomprehension' of the possible futures our technological mastery is propelling us towards 'demands a new sort of thinking', as Nancy puts it, one that is not preoccupied with 'exorcising the purely verbal demon' and 'false concept' which the word 'technology' points towards.[11] The challenge we face today is that technology is revealing itself in increasingly obvious ways neither as a neutral tool to be used well or badly nor simply as a malignant force perverting the human. We live in a time, as Katherine Hayles writes in reference to herself, in which a lack of a phone signal or an internet connection or a low battery can make us 'feel lost, disoriented, unable to work', or even give us the impression that our 'hands have been amputated'.[12] Such feelings of deprivation propel us to the heart of the myth of Prometheus, a myth which suggests that technology is not a 'thing' at all, but that it names a space of possibility which configures the way we think, feel, speak and encounter ourselves and others. Put otherwise, such feelings force us to see something that is otherwise imperceptible: *the trickery of machines at work in the depths of the human soul.*

This book turns to the thought of Günther Anders (born Günther Stern 1902–1992) in order to contribute to Nancy's call for a 'new sort of thinking'. Anders consistently evokes the myth of Prometheus to think through the implications of our technological existence, and a substantial part of this volume is taken up by my translation of his highly provocative essay 'On Promethean Shame'. In this book I suggest that a *progressive step back* to Anders's philosophical anthropology of the technological world, which he developed mainly in *The Obsolescence of Human Beings Vol. 1* (1956) and *Vol.* 2 (1980), can offer highly critical perspectives not only on the world configuring itself today around digital technologies, but also on the manner in which we respond to these generative processes in thought and action.

An engagement with Anders further highlights how the relationship to technological artifice that marks our existence cannot be thought and described directly with the *instruments* of language and reason, for we rely on their trickery even when we mobilise them to advance our insights or attempt to strip the human back to its barest essence. In fact, it is this very dream of mastery that confounds the blindness and limited perspective with which Prometheus sought to protect us from the knowledge of death. Anders can help us see how this blindness is being darkened with ever more fateful consequences.

In the early 1930s, Anders advanced a Promethean definition of the human that stipulates that we are born *unfinished*, as beings who need to make themselves with the help of artifice in order to be at home in the world. Humans are unstable and indeterminate; they exist at a remove from the world:

> To put it paradoxically, *artificiality is the nature of the human beings and their essence instability*. The practical constructions of humanity and its theoretical faculties of representation testify equally to its abstraction. Humans can and must disregard the fact that the world is such as it is, for they are themselves 'abstract' beings; not only part of the world [...] but also 'excluded' from it, 'not of this world'.[13]

This complex quote echoes Plato's *Protagoras* without making a direct reference to it. Anders here plays on the literal meaning of the word 'abstract' – to 'be torn off', drawn away, at a remove (from the world) – and so re-articulates the consequences of Epimetheus's failure to reserve any defining qualities for human beings.[14] Unlike the idealised animals the myth imagines, humans are left 'naked, unshod, without bedding or weapons', which means that they can only grow into the world by turning themselves into new beings with the help of artifice. Read through the work of Anders, the word 'human' merely names the innumerable ways in which we place our hope in artifice to find 'retreats' from our originary lack of power and our exposure to the contingencies of the world. We turn to the trickery of machines to 'retreat' into houses, into thoughts and fantasies that maintain personae that take shape under public scrutiny as well as privately under the covers of our beds; we retreat into clothing and other artificial bodies, harden our hands and expand our reach with tools and instruments; we trick nature into a cultivated landscape that provides marketable goods. In short, our relationship to artifice enables us to retreat into a space of interiority that shields us from the perceived outside that the world constitutes. In so doing, artifice also gives us a space from which to emerge into this very world and play a part in shaping it. Our practical and theoretical constructs and projects thus not only keep us busy and occupy our time, they also fill a void – without them we would not exist, have no being or self to present, find no recourse to meaning. As such, human constructs are only superficially the sign of some positive quality or essence, because they point to an 'estrangedness from the world'. They are the mark of an originary alienation, which Anders calls *Weltfremdheit.*[15]

Because no two ways of retreating from our exposure to the world are identical, because 'my space' is not definitively identical with 'yours', the realm of artifice not only signifies that we are out of touch with the world but also with each other. To have no essence means for Anders that we share an experience of singularity and indeterminacy, but the very experience that separates us energises the artificial bonds that bring us together. Language, for example, opens us to interaction with others as much as it reminds us of our singularity when we feel misunderstood. This estrangement from the world, then, not only gives us a self, it also gives us each other, for it is this exile in artifice that opens us to intimacy and understanding, it enables us – we might be tempted to say – to become 'human'.[16]

This book mobilises Anders's philosophical writings to begin to think through the implications of what it means to find a retreat in the digital and in the spaces opened by machines that can not only activate our ability to think, feel and articulate ourselves, but can increasingly also mimic these very qualities that used to be considered the exclusive domain of the human. Anders's writings on human obsolescence were addressed to a humanity still existing within the 'analog' world of 'non-smart' devices, a world shaped by the formative power of radio, telephone and television. In the first part of this book, the translation of Anders's essay 'On Promethean Shame', digital technology will hence only be an unnamed presence. Despite this absence, the digital moment will nevertheless invoke itself at every turn, because Anders's methodologically exaggerated portrayals of mundane machine interactions practised by a still 'analog humanity' powerfully and uncannily accentuate the hold that digital technologies are increasingly having on all aspects of our lives. Before advancing any further, however, I will set out the intellectual contexts from which Anders's thought emerges and introduce the term *Prometheanism* that will orient the analyses of the ways we find retreats into the digital in the second half of this book.

EXILED AMONG FRIENDS: ON THE PAST AND FUTURE RECEPTION OF GÜNTHER ANDERS

Readers who are familiar with Jean-Paul Sartre's engagements with human freedom, with Martin Heidegger's *Being and Time* or other of his early works, might have been struck by a sense of *déjà vu* while following the account of Anders's early thought given above.[17] This impression of *déjà vu*, or rather, of '*déjà lu*', as Anders describes the niggling feeling of already having read something somewhere,[18] is explained by the circumstance that Anders was a student of Heidegger and Edmund Husserl in the 1920s. Richard Wolin famously called the constellation of now illustrious thinkers who studied under Heidegger at the same time as Anders 'Heidegger's children'.[19] These are: Hannah Arendt, to whom Anders was married from 1929 to 1937, Herbert Marcuse, Hans Jonas and Karl Löwith. It is in reference to this that Martin Woessner calls Anders one of Heidegger's 'stepchildren'.[20] As a prolific writer who published close to 30 books during his lifetime, Anders would no doubt deserve to be named among his more canonical contemporaries.[21] The term 'stepchild' certainly also records the influence of Heidegger's thought that is evident across Anders's body of works. It is, however, not hard to imagine how Anders would have responded to any close association with his former teacher, let alone a quasi-familial one. While introducing Volume 2 of *The Obsolescence of Human Beings* in 1979, Anders recounts a conversation he had held fifty years earlier with a 'nowadays world famous philosopher who is as morally impoverished as he is brilliant in theoretical thought'. The reference clearly is to Heidegger, and Anders relates how in 'his own peculiar manner of taking delight out of being contemptuous' Heidegger had warned him against '*ever deserting into responding to practical matters*'.[22] As such a tendency to desertion already needs to be evident before a warning against it can be uttered, it will come as no surprise that Anders most emphatically refused to heed to this advice.[23] Anders, much more so than any of his philosophical contemporaries, is a pragmatic thinker whose writings force us to actively confront and disrupt the feelings of everydayness and normality which ossify an artificially produced human world into a seemingly natural state of affairs. 'The point of departure and object of all my reflections', as Anders puts it, 'are isolated and entirely concrete phenomena (*ganz konkrete Einzelphänomene*) of our contemporary life'.[24]

Anders's thought, and this is particularly evident in the terminology of his pre-war writings on the 'human's estrangement from the world', is marked by Heidegger's fundamental ontology and the divergent phenomenological approaches of Husserl and Max Scheler (for whom Anders worked as an assistant in the late 1920s). In the introduction to his 1984 book *The Human without World*, Anders elaborates how this idea took shape during a phase of his life that constituted a 'totally apolitical intermezzo', a theoretical attitude that had become 'incomprehensible' in the course of his lifetime.[25] As a secular Jew who fled Nazi Germany for Paris in 1933 and then on to California in 1936 to commence an American exile that lasted until 1949, his life was marked by '*practical matters*' in ways that remained inconceivable to his former teacher and much of the intellectual and cultural environment of which he was part in Germany.[26] Anders's pragmatic approach to conducting philosophy; his reliance on experiences made while working in factories, movie repositories and in other odd-jobs; his ardent political activism, as well as his style of writing need to be viewed against the backdrop of the failure of the academic establishment to mount any sustained resistance to the rise of National Socialism.[27] Anders saw this academic failure replicated on the world stage after the detonation of the first nuclear bomb over Hiroshima on 6 August 1945. The following remark, lifted from a 1987 interview, succinctly encapsulates both the stylistic and thematic trajectory of Anders's writings after leaving Germany: 'It seemed nonsensical and comic to me, if not even downright immoral, to produce texts about morality that could only be read and understood by academic colleagues. As nonsensical as if a baker would only bake bread-rolls for other bakers.'[28] As will become evident in the course of this book, this commitment to accessibility and 'practical matters' by no means leads to one-dimensional or impoverished thinking, but to a unique style of writing that is as philosophically profound as it is politically and morally incisive.

Konrad Paul Liessmann, who for decades has been a leading expert and prolific contributor of German language scholarship on Anders, aptly sums up Anders's recalcitrant position by calling him 'Philosophy's Outsider'. After stepping outside of Heidegger's and Husserl's approaches to philosophy, his engagements with National Socialism shared a proximity with Marxism and the Frankfurt School of Critical Theory, but pursued their own course, and led to 'a fragile existence' as a freelance author rather than an academic career. 'He also', as Liessmann elaborates, 'remained an outsider wherever he lived: As a Jew in Germany, as a European in America, as a remigrant in Austria [...] as an atheist in Judaism'. These often self-cultivated positions of intellectual, institutional, national and ideological exile set the stage for the position that is perhaps most defining of all. Anders seems to be outside of time. Although profoundly marked by the events of the twentieth century, his writings had – and continue to portray – a tendency to isolate specific themes and developments *earlier* than his academic contemporaries. He wrote on television, space-flight and especially on the terrible logic of nuclear weapons, while popular, political and academic discourses had either 'still not acknowledged these technological developments or were celebrating these triumphantly', as was the case with the advent of the nuclear age.[29] 'It is one of my vices that I always arrive too early', Anders suggests, reflecting on the untimeliness of his thinking in a remark to Liessmann, 'and whoever is too early does not arrive on time either'.[30]

This leads me to another sense in which Anders can be seen to be 'exiled amongst friends'. His name appears in the margins of some of the most influential contemporary critical responses to the ways in which technological developments are reconfiguring our existence, for example in the work of Zygmunt Bauman, Thierry Bardini, Jean-Pierre Dupuy, Veit Erlmann, Roberto Esposito,

Maurizzio Lazzarato, Jean-Luc Nancy and Bernard Stiegler.[31] With the exception of *The Life and Work of Günther Anders*, and pioneering essays by Babette Babich, Jason Dawsey, Michael Hauskeller and Ernst Schraube, detailed engagements with Anders's critique of technology are still missing in Anglophone scholarship, in relation to twentieth-century continental thought in general, as well as in the specific context of what David Wills and Arthur Bradley have called the 'technological turn' in contemporary thought.[32] As Wills elaborates in *Dorsality*, the challenge to which this turn to technology responds consists in 'taking conceptual account of the extent to which, in increasingly explicit ways, technology defines and redefines the human and does so downstream from the point at which a given technological creation was brought into effect'.[33] It is this problematic that has oriented my own engagement with Anders's 'Promethean' conception of technology, and as such I will introduce Anders's central ideas through an engagement with the thinkers of the 'technological turn', who frequently build on Stiegler's work. As Stiegler has shown in *Technics and Time, 1*, the Prometheus myth offers a highly productive path to thinking the relationship between humanity and technology, because it makes the 'invisibility' of technology's formative effect visible, because it presents the human not as a tool user but as a technological being, a being that exists because of its relationship to technology as much as being continually re-shaped by this relationship.

One reason why Stiegler mobilises the myth is to illustrate and philosophically develop André Leroi-Gourhan's pioneering paleoanthropological account of human evolution, the logic of which feeds into much of the work that makes up the 'technological turn' in contemporary thought. Viewed from the paleoanthropologist's perspective, the skeletal remains of the first anthropoids found amidst most primitive tools point towards a process of human and technological co-evolution. For Leroi-Gourhan, the human does not evolve to eventually make use of tools but evolves *because of* tool use. What the myth presents as the symptom of an originary defect thus for Leroi-Gourhan marks an evolutionary advantage that the human has over other forms of life. In *Gesture and Speech*, the increasingly upright stance that emerges in the course of human evolution gradually frees the hands for tools, the mouth for speech and then finally the brain for 'human' thought, conscience and sociality:

> Freedom of the hand almost necessarily implies a technical activity different from that of apes, and a hand that is free during locomotion, together with a short face and the absence of fangs, commands the use of artificial organs, that is, of implements. Erect posture, short face, free hand during locomotion, and possession of movable implements – those are truly the fundamental criteria of humanity. [. . .] The fact that my list does not include a large brain may be thought

> surprising. Of course it is difficult to assign pre-eminence to any particular characteristic, since in the development of species everything is interlinked, but I believe that there can be no doubt that to some extent cerebral development is a secondary criterion. Once humanity has been achieved, the brain plays a decisive role in the development of human societies. In terms of strict evolution it is undoubtedly a correlative of erect posture and not, as was thought for a long time, primordial.[34]

Technology is here not seen as the endpoint of human evolution but as its beginning. In *Technics and Time, 1*, Stiegler succinctly sums up this process by calling the history of technology '*the evolution of the living by other means than life*'.[35] For Stiegler, this means that '"human nature" consists only in its technicity, [only] in its denaturalisation'.[36] And with this notion we have, in effect, arrived back at our point of departure: Anders's suggestion that '*artificiality is the nature of human beings and their essence instability*'.

If humanity is understood to be continually redefined and reshaped through its exposure to artifice that is itself created by humans, then it is not only technology that reveals itself as a 'fetish word', but general descriptions of humanity also. For 'our' humanity takes shape only as we take the step into artifice, only as we grow up into the artificial structures that pre-exist us and continue to evolve in the course of our lives. To be artificial by nature means, as Anders puts it, that '*the human is general in a very specific* [*spécial*] *way*; it is not realised according to a single form foreseen in principle and valid in general, but, as daily life and history indicate, according to many different types. […] It is the fact of the variation, and not the constancy of the variable, that defines the specifically human […].' To speak of human beings in the plural as 'humans' thus presumes and fetishises an '*a priori* humanity' which is general only insofar as it is 'indeterminate', shaped through a relationship to artifice that plays itself out in an irreducibly singular way in each human being.[37] Language, perhaps the most intuitive skill that Plato ascribes to Prometheus – humans soon used 'their skill to develop articulate speech and words for things'[38] – is a prime example of how a *shared* capacity and linguistic indeterminacy (we can learn any language) nevertheless create irreducible *singularities*, that is, a plurality of distinct speakers. These speakers might have English, French, German or further languages in common to varying degrees, but this shared medium brings speakers together only by also accentuating and shaping their singularity. The shared capacity to speak and embody artifice not only produces singular ways of speaking (accents, inflections, voices), it also brings out singular stories, and allows 'shared' words to be continually re-inscribed and coloured by experiences that are registered in a medium that acts as a boundary as much as it is a gateway. What I understand a particular word to mean, the feelings and thoughts I associate with it, can never be fully disclosed. Far from being a 'prison house', then, language opens us to an expression and experience of self rather than limiting it.

This small illustration should suffice to highlight why a form of thinking is required that does not fall back onto a logic that seeks to develop its position by relying on general or unquestioned preconceptions and assumptions about the 'human' or 'technology', or indeed about the manner in which these two 'fetish words' relate. For if it is understood that 'technology' is not the name for a thing, but rather for a generative process that is intertwined with life from the beginning, then thought seems to encounter in technology and in the question of artifice an intrinsic limit beyond which it cannot step: 'A technology that defines and so produces the human, cannot be part of the human self-image.'[39] Wills here highlights an originary blindness towards the technical that thought cannot directly expel. If technology is understood to always be 'in the back' of the human – if it propels us from behind into an upright stance and into a future that we cannot fully grasp – then we need to consider a relationship between humanity and technology which, as Wills puts it, remains 'inaccessible or invisible from the perspective of an integral human gathered within its neatly prescribed limits or borders and gazing ahead into a controlled exteriority of the artefact'.[40] Technology, to use Wills's evocative notion of 'dorsality', is a blind spot in our back: it invariably turns away from us the moment we seek to turn towards it in thought and feeling. If we take into account, therefore, that a relationship to technology has not only shaped 'the' human to the extent that the fabric of the human body and brain has evolved alongside the technological, but that a relationship to artifice also shapes the self-consciousness and self-awareness that each body animates in an irreducibly singular fashion, then thought needs to begin to think through the effects of this blindness rather than seeking to expel it by producing illuminating insights and general definitions. For how could we hope to isolate and respond to the formative effects and evolution of technological artefacts and the signifying machines we mobilise in the course of our lives, if these have already shaped the way we apprehend our own existence? The various approaches to thinking the originary interrelation of life and technology that make up the 'technological turn' in theory and philosophy hinge on this question. Or so I suggest.

'Prometheanism', a term Anders deploys in 'On Promethean Shame', and which I have mobilised as the title of this book, seeks to capture this movement towards a new form of thinking about technology in the wake of Heidegger, Leroi-Gourhan and the unfolding of a technological modernity in which the 'unbridled character of our mastery' is manifesting itself in increasingly visible ways. With this term, therefore, I do not seek to categorise a group of thinkers, or engage primarily with its recent uses in the context of 'Accelerationist' responses to the current state of politics.[41] Rather, I wish to frame the following problematic: if the word 'technology' cannot simply name *the object* that our contemplative gestures seek to get hold of, but must rather be understood as naming a relationship to artifice that opens us to *the very possibility of thought*, then how can we respond to concrete and tangible technological developments? How can we think about technology and the concrete effects machines have on our lives without falling back into the pattern of simply opposing humanity to a technological other? Over recent decades these questions have given rise to exciting and highly productive new ways of thinking the proliferation of machines and the new constellations and concentrations of power they herald.[42] The central thesis of this book is that Anders can make a contribution to this field of enquiry that has been overlooked for too long. Much of his thought – and the essay included in this volume is a case in point – hinges on momentary feelings of curtailment, deprivation and obsolescence that become manifest while we use machines. These fleeting and often unacknowledged emotive tensions and resistances are seen as symptomatic of a relationship to artifice that is so intuitive that the formative effects that artifice has in general are not noticed at all.

Anders mobilises his phenomenology of machine use in order to illuminate the invisibility and feeling of normalcy and everydayness that naturalise artificial structures. Machine processes happen without our (emotional or physical) involvement: they take care of things in our stead and as such they shape the way we relate to ourselves because they slip from our consciousness. These processes are – if an analogy is permitted – as invisible and empowering as language is when it intuitively offers itself as an instrument at our disposal. That is to say, artificial media remain invisible instruments only while they are complicit; as Heidegger puts it with regard to language, it is invisible only 'until we cannot find the right word for something that irritates or gives us joy'.[43] For Anders, machine-artifice configures our soul, which makes us feel all the more worried if we are shown up as inadequate machine users or inarticulate writers. It is in these moments of emotional turmoil and helplessness that the formative power of artifice reveals itself only to remind us of our innate lack of ability, powerlessness and obsolescence. These feelings lead Anders to a new conception of human finitude based around our inability to see or comprehend the artificial powers we blindly place our hope in. Every technological innovation has unforeseen effects and consequences while also reconfiguring the parameters of human existence. The further technology advances, the more it impresses on us that the human, as it is born, is obsolete, useless and embarrassing. As humans we are born *weltfremd*, strange to a world we can only inhabit by retreating into artificial structures. This very retreat into artifice, Anders suggests, is increasingly also leading to a world without humans, a world in which human existence is becoming ever more conditional and exposed to the power of artifice that highlights and accentuates every human flaw.

By isolating moments of curtailment and feelings of dependency while using machines, a central concern of Anders's thought is to make visible how our relationship to artifice is continually reconfiguring our ethical and moral sensibilities, in short, our feelings. In *Ketzereien* (*Heresies*), Anders calls his own approach '*philosophy of discrepancy*'.[44] The term 'discrepancy' captures the key concept that crystallises in the course of his post-war writings, the notion of a 'Promethean slope or gradient' (*prometheisches Gefälle*), a growing rift between our technologically mediated ability to collectively influence the world and our individual capacity to feel, and to emotively apprehend, what we are doing. As this is put in 'Reflections on the H Bomb', one of the few texts available in English: 'The performances of our heart – our inhibitions, fears, worries, regrets – are inversely proportional to the dimensions of our deeds, i.e., the former grow smaller as the latter increase in size.'[45] With this formula Anders describes the tendency of machines to generate, maintain and increase an emotional distance between human actions and their effects. The atom bomb – a technological object that can project power 'without us' merely by existing – relieves an administering power of the need to motivate human beings to exert the emotive effort required to threaten others or even to personally annihilate a city. Once the bomb exists, human action can be focused solely on maintaining the bomb, for it extends a threat by its mere existence. This example, to which I return in detail in the second part of this volume, gives a first impression of what, in loose analogy to Susan Buck-Morss's work on Benjamin, I call the *anaesthetic force and violence* of technological objects.[46] With this term I seek to capture a central motif of Anders's thought: the absentmindedness and absent-heartedness that automatically configures itself around complex machines. 'It would be short-sighted', as Anders elaborates, to deem the bomb exceptional in this respect, for it rather reveals the 'principle' every machine is 'endowed' with: 'What we aim at with machines is to produce an effect that does not require our presence or help and unfolds without lament and complaint (*klaglos*).'[47] In short, the *anaesthetic force* Anders discovers names the capacity of machines to relieve us not only of pain and suffering, but also of worry and existential concern.

This anaesthetic tendency can help us recognise how this outsourcing of hope compounds an originary blindness towards the artifice that shapes our perception of self, a blindness which is continually being re-modulated, and the effects of which are being amplified. This blindness reaches critical levels, or so Anders suggests, just as technology provides us with ever deeper insights into the structures of our physiology and the universe, awakening the dream that we can finally 'see' what we are ourselves. Increasingly autonomous and complex machines have the capacity to blunt and reconfigure our feelings and sensibilities while also exposing us to moments of uncomfortable emotional feedback: The very machines we place our hopes in begin to channel and amplify the concerns we had hoped to outsource with their help. 'I stopped mortals from brooding on death', 'I made to lodge within them blind hopes', as Aeschylus has the Titan proclaim – yet by turning us away from our limitations and finitude, our Promethean trickery increasingly also accentuates the flaw. With this we have arrived at the central motif of this book: Günther Anders's notion of 'Promethean shame', the shame in view of the 'humiliatingly high quality of manmade objects'. One of the ways in which this shame manifests itself is in the manner it reconfigures the way we experience our own finitude, the very relationship to death that Aeschylus (and a long tradition of thinkers in his wake) deems constitutive of human existence.

Anders suggests that the more invisible and abstract the act of dying becomes in technologically advanced societies – the more we can be cured of disease and the effects of ageing, the more death happens on TV screens in parts of the world we do not need to care about, the more distant we are from death and the other's dying – the more mortal and vulnerable we seem to become. The ever better cure draws attention to finitude and limitation in ever more powerful ways. For Anders the original turn away from death is 'exclusively of the order of negation': we turn to artifice because '*we do not want to die*', but this does not mean that 'we want to continue to live on and on', or that we are capable of maintaining a 'positive mental image' of such an eternity, of an absolute turn away from finitude. 'Our imagination', as Anders puts it, 'stays *blind*' in view of such an eternity because we are '"too mortal" to even positively "*deem*" that we will never die'.[48] By reconfiguring our limitations with the help of artifice we do not remove ourselves from an existential condition; we merely unleash and amplify the desire to negate finitude, the very desire that binds us to it. The spotlight of shame turns such limitation into a pathological failing and is experienced as a disease that needs to be concealed. Amidst highly capable machines, Anders suggests, being a mortal human becomes a shameful flaw, and ultimately a condition we hope to be absolved from. Put otherwise, the more limitless machine ability becomes, the more limited and finite we feel, and the more anxious we become. The anaesthetic force of technological objects, their capacity to remove us from our body thus coincides with a hypersensitivity towards our own individual limitations and exaggerates our awareness of our own potential flaws. As Michael Hauskeller has shown in a brilliant reading of 'On Promethean Shame' to which I return below, the human enhancement project and the obsessive self-tracking that smartphone apps enable, can be read against the backdrop of the very feeling that Anders isolates. Put otherwise, they are motivated by a new form of bodily shame, 'the shame of having been born rather than having been made'.[49]

This short glimpse ahead into 'On Promethean Shame' gives an indication of how Anders's writings anticipate the call for a new sort of thinking that

I have mobilised through the words of Jean-Luc Nancy. In doing so they help us appreciate that in order to understand the technological, we need to consider our experience of finitude, and the way this experience is itself configured by artifice, rather than fetishising the digital moment as an inherently 'new' phenomenon. One of the main strengths of Anders's writings, as I see it, is their ability to switch registers *from* existentialist-phenomenological observations of how machines shape and reconfigure our feelings and perception by functioning 'without us' *to* critiques of the modes of production and concentrations of power that accompany specific technological advancements. Anders's thought opens a perspective on our digital future that can help us see how machines ever more efficiently sanitise conditional spaces and constellations of power that are disclosed in the course of our technological modernity. In the world that is taking shape, *Weltfremdheit*, the originary 'estrangedness from the world' that makes us retreat into artifice, concretely means: 'not being granted admittance to the world' (*zur Welt nicht-zugelassen-worden-Sein*).[50] In this context, then, *Weltfremdheit* points towards a world in which the vast majority is blocked from any mobility in the same way that cars are kept from the road if they are not deemed 'road legal', that is, are not 'registered as roadworthy' (*zugelassen*). 'Being-without-world' (*weltlos*) thus also describes the most concrete, 'class specific' and economic reality of material property. This means concretely, as Anders elaborates, 'that the world that a human being is creating, or helps to create, does not belong to *this* human being, who is at home in the world as little as the builder is at home in the building that he helps to erect for someone else'.[51] It is to this problematic that I return in the second part of this book which focuses on the anaesthetic force of technology that Anders allows us to think. Read through Anders, the debates currently revolving around artificial intelligence and digital technologies show themselves to be a supplementary ruse, for humans are not in danger of becoming obsolete because they might be replaced by intelligent machines, the innate obsolescence that has marked humanity from the beginning is 'merely' being amplified with ever more fateful consequences. As machine processes are increasingly capable of relieving us not just of physical but also of emotional and cognitive effort by allowing us to automate our correspondence and acts that we might not have the heart to do, a cold, calculative, non-human world is being established under the cover of a deceptive sense of normality. To develop this perspective I turn to the phenomenon of 'unsalaried work', which I define as the work we conduct to make ourselves obsolete, expendable and ultimately inhuman.

As I have already indicated above, the principal aim of this book is to contribute to the new chapter in the reception of Anders in the English speaking world, a chapter that has long since been opened in Germany, France and Italy.[52] In the late 1980s, at the vanguard of a now-vast German language scholarship and philosophical reflection on Anders's work, Gabriele Althus complained that 'one can read much *by* Günther Anders, but little *about* him'.[53] In English speaking contexts this statement would have to be modified to say something along the lines that 'one can read a little about Anders but almost nothing by him'.[54] Anders's highly fascinating life-story, and his many close bonds with thinkers and intellectual figures that have shaped twentieth-century thought – his cousin Walter Benjamin, his first wife Hannah Arendt, Bertold Brecht, Sigmund Freud, Theodor Adorno, to add a few names to the ones already mentioned – have great power to distract from the originality of his own thought and the contribution it can make to our understanding of the technological reality we are part of today. It is for this reason that the perspective this book opens onto our digital moment branches out from a translation of the substantial essay 'On Promethean Shame'. The essay was first published in 1956 as part of *The Obsolescence of Human Beings 1: On the Soul in the Age of the Second Industrial Revolution.* 'On Promethean Shame' is the first of four essays that make up the volume.[55] It isolates feelings of inadequacy, panic and obsolescence that can befall us amidst ever more capable machines. The discussions of a wide range of cultural phenomena and concrete technological objects (ranging from '*Iconomania*', the addiction to images taken of oneself, to surveillance, bodily enhancement and changing working patterns) are framed by a highly original phenomenological analysis of the feeling of shame itself and by a discussion of the manner in which technological advancement is changing our relationship to finitude. Because the technological environment Anders is responding to has evolved, and is continuing to do so at an accelerating rate, his exaggerated portrayal of the mundane world of a still 'analog' humanity powerfully and uncannily accentuates the hold that digital technologies have on us today. It is this opening that I pursue in the second half of this volume, which I will introduce separately below. To lead into the essay 'On Promethean Shame', I will simply relay the 'warning' with which Anders introduces '*philosophical exaggeration*', his preferred method of thinking, at the outset of *The Obsolescence of Human Beings 1*: 'At least some of the portrayals advanced here will seem to be "*exaggerated.*" This impression is easily explained: they *are* "exaggerated."'[56]

NOTES

Note on Translation: Unless stated otherwise, translations from works by Günther Anders are my own. I generally use my own translations of German book and essay titles in the text (but not in the notes, where I reproduce the original German sources). Possible translations of titles are given in the bibliography.

1. Jean-Luc Nancy, 'A Finite Thinking', *A Finite Thinking*, ed. Simon Sparks, trans. Edward Bullard, Jonathan Derbyshire and Simon Sparks (Stanford: Stanford University Press, 2003), 3–30, 25.

2. Günther Anders, 'On Promethean Shame', (ch1) 31* below.

3. Hesiod, *Theogony and Works and Days*, trans. Catherine M. Schlegel and Henry Weinfield (Ann Arbor: University of Michigan Press, 2006), 39.

4. Bernard Stiegler, 'Prometheus's Liver', in *Technics and Time 1: The Fault of Epimetheus*, trans. Richard Beardsworth and George Collins (Stanford: Stanford University Press, 1998), 185–203, 186.

5. Plato, 'Protagoras', in *Gorgias, Menexenus, Protagoras*, ed. Malcolm Schonfield, trans. Tom Griffith (Cambridge: Cambridge University Press, 2010), 157.

6. Aeschylus, *Prometheus Bound*, trans. A. J. Podlecki (Oxford: Aris & Phillips, 2005), 93.

7. Pierre Hadot, *The Veil of Isis: An Essay on the History of the Idea of Nature*, trans. Michael Chase (Cambridge, MA: Harvard University Press, 2006), 94. Hadot offers this definition in the context of outlining the 'Promethean attitude' towards nature, which consists of the desire to unveil and master nature's secrets with the help of technology. See especially 91–155.

8. Hesiod, *Theogony and Works and Days*, trans. M. L. West (Oxford: Oxford University Press, 1999), 38.

9. See Erik Brynjolfsson and Andrew McAfee, *The Second Machine Age: Work, Progress and Prosperity in a Time of Brilliant Technologies* (New York: W.W. Norton & Company, 2014). Brynjolfsson and McAfee elaborate on this with unbridled Promethean optimism: 'We are convinced that we are at an inflection point – the early stages of a shift as profound as that brought on by the Industrial Revolution. Not only are the new technologies exponential, digital, and combinatorial, but most of the gains are still ahead of us. In the next twenty four months, the planet will add more computer power than it did in all previous history. Over the next twenty four years the increase will likely be over a thousand fold. […] Our generation will likely have the good fortune to experience two of the most amazing events in history: the creation of true machine intelligence and the connection of all humans via a common digital network, transforming the planets economics' (p. 251).

10. Arthur Bradley, *Originary Technicity: The Theory of Technology from Marx to Derrida* (Basingstoke: Palgrave MacMillan, 2011), 5.

11. See Nancy, 'Finite Thinking', 25.

12. Katherine Hayles, *How We Think: Digital Media and Contemporary Technogenesis* (Chicago: University of Chicago Press, 2012), 2.

13. Günther (Stern) Anders, 'The Pathology of Freedom: An Essay on Non-Identification', trans. Katharine Wolfe, *Deleuze Studies* 3 (2009), 279. [Translation modified]

14. I am here following Werner Reimans's reading of the meaning of the word 'abstract' in Anders's thought. See 'Nihilismus und Scham', in *Günther Anders Kontrovers*, ed. Konrad Paul Liessmann (München: C.H. Beck, 1992), 57–71, 61.

15. See esp. Anders, 'Pathology of Freedom', and 'Une interprétation de l' a posteriori', trans. Emmanuel Levinas, *Recherches Philosophiques* 4 (1934–1935), 65–80.

16. In the course of an 1930 essay on Rilke's *Duino Elegies*, Anders and Hannah Arendt, who were (still) married at the time, offer the following elaboration on the intrinsic connection between love, singularity and *Weltfremdheit*: 'Negatively expressed, being estranged from the world means to find oneself excluded from the world. In love, *Weltfremdheit* finds its positive manifestation as a being "unbound and open for someone"' (p. 868). See Hannah Arendt and Günther Stern (Anders), 'Rilkes *Duineser Elegien*', *Neue Schweizer Rundschau* 11 (1930), 855–71.

17. For a discussion of Anders's early work, its origins and influence see Eckhard Wittulski, 'der tanzende Phänomenologe', in *Günther Anders Kontrovers*, ed. Konrad Paul Liessmann (München: C.H. Beck, 1992), 17–33.

18. Anders, 'On Promethean Shame', (ch1) 44* below.

19. See Richard Wolin, *Heidegger's Children* (Princeton and Oxford: Princeton University Press, 2001).

20. See Martin Woessner, *Heidegger in America* (Cambridge: Cambridge University Press, 2011), 67.

21. A list of most of Anders's main philosophical writings and some, but by no means all, secondary works in German can be found in the Bibliography. Besides his philosophical writings, Anders also published several novels and literary narratives and hundreds of poems. Heinz Schaeffelmeier has compiled a comprehensive bibliography of Anders's works to which I would like to refer the reader. In addition to a full list of books published during Anders's lifetime and the growing number of posthumous book publications, Schaeffelmeier's bibliography records the many hundreds of shorter publications (book chapters, journal articles, published speeches, poems, interviews and journalistic pieces) that Anders produced in the almost seventy years in which he was active as a publishing writer. Schaeffelmeier also provides a comprehensive overview of the repository of Anders's unpublished writings and correspondences, housed by the *Literaturarchiv* of the Austrian National Library in Vienna. The bibliography also includes an overview of Anders-scholarship and existing translations. Web link: http://www.forvm.at/texte/ga_bibliographie.html. Harold Marcuse's excellent webpage on Günther Anders also includes a comprehensive bibliography and provides links to pdfs of translated work: http://www.history.ucsb.edu/faculty/marcuse/anders.htm. The webpage of the International Günther Anders Society provides up to date information on publications, events and information about ongoing editorial projects and posthumous publications: http://www.guenther-anders-gesellschaft.org.

22. Günther Anders, *Die Antiquiertheit des Menschen 2: Über die Zerstörung des Lebens im Zeitalter der dritten industriellen Revolution* (München: C.H. Beck, 2002), 12.

23. In 1988, Anders succinctly summed up his relationship to Heidegger as follows: 'Even though I still believe today that certain analyses in *Being and Time* are the most profound philosophical writing this century has seen, I feel morally more removed from Heidegger than from any other Philosopher.' See '*Unverständlich*', a short newspaper announcement in which Anders rebuts Michael Haller's suggestion that Anders's philosophy of technology was influenced by Heidegger's post-war writings. *Die Zeit* 18 (1988).

24. Anders, *Antiquiertheit* 2, 414.

25. Günther Anders, *Mensch ohne Welt: Schriften zur Kunst und Literatur* (München: C.H. Beck, 1993), xiv.

26. The realities of life as an émigré are a central aspect of Anders's thought. For a profound reflection, see, for instance, 'The Émigré', which has recently appeared in translation in *The Life and Work of Günther Anders*, eds. Günther Bischof, Jason Dawsey and Bernhard Fetz (Innsbruck: Studienverlag, 2014), 171–86.

27. Anders illustrated the apolitical stance of his academic environment by imparting how he had been ridiculed by his colleagues when he had tried to organise discussion groups on Hitler's *Mein Kampf* in 1928. See Konrad Paul Liessman, *Günther Anders: Philosophieren im Zeitalter der technologischen Revolutionen* (München: C.H. Beck, 2002), 21.

28. Elke Schubert (ed.), *Günther Anders Antwortet: Interviews und Erklärungen* (Berlin: Tiamat, 1987), 28. The image is extended as follows: 'If bakers only baked bread for bakers, if tailors only clothed tailors, dentists only filled the teeth of dentists, then they would hardly be acting more esoterically than university professors are today' (p. 182).

29. I am here summarising Konrad Paul Liessmann, 'Between the Chairs: Günther Anders – Philosophy's Outsider', in *The Life and Work of Günther Anders*, eds. Günther Bischof, Jason Dawsey and Bernhard Fetz (Innsbruck: Studienverlag, 2014), 73–83, 73*ff.* In addition to Liessmann's essay, which I cite here because it is easily available in English, I have mainly relied on Jason Dawsey's excellent introduction to the *Life and Work of Günther Anders* and the following German language introductions and biographical texts: Liessmann's, *Günther Anders zur Einführung* (Hamburg: Junius, 1988), and *Günther Anders: Philosophieren im Zeitalter der technologischen Revolution* (Munich: C.H. Beck, 2002), and Elke Schubert's *Günther Anders* (Hamburg: Rowohlt, 1992).

30. Liessmann, *Günther Anders zur Einführung*, 12.

31. The versatility of Anders's mode of questioning is reflected in the wide thematic range of recent engagements. See for instance, Thierry Bardini, *Junkware* (Minneapolis: University of Minnesota Press, 2011); Zygumnt Bauman, *Consuming Life* (Cambridge: Polity, 2007); Jean-Pierre Dupuy, *The Mark of the Sacred*, trans. M. B. Debevoise (Stanford: Stanford University Press, 2013); Veit Erlmann, *Reason and Resonance: A History of Modern Aurality* (Cambridge, MA: MIT Press, 2014); Robert Esposito, *Immunitas*, trans. Zakiya Hanafi (Cambridge: Polity, 2013); Bernard Stiegler, *What Makes Life Worth Living: On Pharmacology*, trans. Daniel Ross (London: Polity, 2013); Jean-Luc Nancy, *After Fukushima: The Equivalence of Catastrophes*, trans. Charlotte Mandell (New York: Fordham University Press, 2015). These texts have in common that they are written by authors who had access to French and Italian translations or were able to consult German originals. I suggest that 'revisiting' Anders's thought will reveal a philosophical, critical and political potential that will strongly outweigh some of its inevitable shortcomings.

32. See Bradley, *Originary Technicity*, 2; David Wills, *Dorsality* (Minneapolis: University of Minnesota Press, 2008), 3. Bibliographical information of engagements with Anders in English is provided on citation.

33. Wills, *Dorsality*, 7.

34. André Leroi-Gourhan, *Gesture and Speech*, trans. A. Berger (Cambridge, MA: MIT Press, 1993), 19.

35. Stiegler, *Technics and Time 1*, 135.

36. Stiegler, *Technics and Time 1*, 148. In 'The Technics of Decision' this notion is further explained in the following broad terms: 'Now, by manufacturing these artefacts – for example, in order to cut meat, to cut down trees, to work materials, to fight – this new form of life – [Stiegler is talking about the human] – creates a new basis for memory. Every living sexual being is constituted by two kinds of memory which, since August Weismann, have been called germinal and somatic – the genetic memory of the species preserved in DNA, and the somatic, individual memory preserved in an organism's nervous system. Now, with what Leroi-Gourhan calls the process of exteriorisation, there appeared around three million years ago a third form of memory: *technical* memory that enables the transmission of the individual experience of people from generation to generation, something inconceivable in animality.' See 'Technics of Decision an Interview', trans. Sean Gaston, *Angelaki* 8.2 (2003), 158–9.

37. See Anders, 'Pathology of Freedom', 286. Anders would later berate himself for the traces of essentialist anthropocentrism that this formulation still contains. See 'On Promethean Shame', (ch1) 89* below.

38. Plato, 'Protagoras', 158.

39. Wills, *Dorsality*, 7.

40. Ibid.

41. I engage with 'Accelerationist' politics in the second part of this book.

42. Besides the approaches already mentioned, which broadly speaking unfold in the wake of Heidegger and Derrida (Bradley, Nancy, Wills, Stiegler), I am here specifically thinking of: Italian *Operaismo* (and especially Matteo Pasquinelli's responses to it) among a wide range of further Marxist perspectives on the emerging attention economies; Catherine Malabou's pioneering work on plasticity; the thought of Bruno Latour, Jaron Lanier and Sherry Turkle, as well as the multifaceted lines of enquiry that engage with Posthumanism and Affect Theory.

43. See Martin Heidegger, *Unterwegs zur Sprache* (Stuttgart: Klett-Cotta, 2007), 161 (my translation). I return to the 'negative perceptibility' of the structuring power of artifice at length.

44. In Anders's own words: 'The editor of a dictionary asked me how he should classify my "philosophy." He told me that he was hesitant to simply add it to "Phenomenology," "Existentialism," "Marxism" or even "Critical Theory." He was right to hesitate. […] I have authorised the term "*Philosophy of Discrepancy*" ("*Diskrepanzphilosophie*"). For the unprecedented point of departure which all of my analyses share […], is the discrepancy that exists between the various capacities of human beings, a discrepancy that means that we can produce more than we can imagine. [I told the editor] that my books – not only the two volumes of *The Obsolescence of Human Beings* – portray the effects of this discrepancy.' See *Ketzereien*, 225.

45. Anders, 'Reflections on the H Bomb', trans. Norbert Guterman, *Dissent* 3.2 (Spring 1956), 146–55, 154. Translation modified.

46. See Susan Buck-Morss, 'Aesthetics and Anaesthetics: Walter Benjamin's Artwork Essay Reconsidered', *October* 62 (1992), 3–41.

47. Günther Anders, *Die atomare Drohung: Radikale Überlegungen zum atomaren Zeitalter* (München: C.H. Beck, 2003), 199.

48. See 'On Promethean Shame', (ch1) 57–58*. It is worth pausing here. Anders's thought emerges from what might be called a 'tragic' tradition of thought that sees in our incomprehension of the fact that we will die the constitutive blindness of human existence. In addition to Heidegger's discussion of 'being-toward-death' (See Sections 45–53 of *Being and Time*), we might consider the following succinct and influential passages. Immanuel Kant: 'No human being can experience his own *death* (for to constitute an experience requires life), he can only observe death in others.' See 'Anthropology from a Pragmatic Point of View', trans. R. Louden, *The Cambridge Edition of the Works of Kant: Anthropology, History, and Education* (Cambridge: Cambridge University Press, 2007), 277. Sigmund Freud: 'It is true that the proposition "All men are mortal" is paraded in text-books of logic as an example of a generalization, but no human being really grasps it, and our unconscious has as little use now as ever for the idea of its own mortality. Religions continue to dispute the undeniable fact of the death of each one of us and to postulate a life after death; civil governments still believe that they cannot maintain moral order among the living if they do not uphold this prospect of a better life after death as a recompense for earthly existence.' See 'The Uncanny', trans. Alix Strachey, *The Standard Edition of the Complete Psychological Works of Sigmund Freud Volume XVII: An Infantile Neurosis and Other Works* (London: Hogarth, 1955), 242. Kant's, Freud's and Heidegger's articulations of this formative blindness can all be read as modulations of the blindness towards death that Aeschylus's tragedy pinpoints. As such, *Prometheus Bound* introduces one of the central motifs of the tragic, a genre that can be understood as a 'machine' that aims to 'trick us' into seeing our own blindness. In doing so the tragic (as well as the philosophical tradition that follows in its wake) leave us with an impenetrable paradox: do we become more insightful when we understand our own blindness towards death, or does our understanding replicate the error to which we see tragic heroes such as Prometheus and Oedipus fall victim to – for does the tragic not automatically lead us back to a false impression of insight thus providing us with new grounds for hope? It is this *aporia* that shapes Anders's thought. I return to this question in the second part of this book, while engaging with the 'transhumanist' desire to negate death.

49 See Anders, 'On Promethean Shame', (ch1) 30*.

50. Anders, *Mensch ohne Welt*, xiii.

51. Ibid., xii.

52. For an extensive bibliography of recent translations and international scholarship I would here like to refer the reader to Jason Dawsey's introduction to *The Life and Work of Günther Anders*, esp. 12–14.

53. As cited by Dawsey in *The Life and Work of Günther Anders*, 12.

54. In view of the originality of Anders's thought and its current resurgence abroad, the lack of English language translations seems baffling and is, as Babette Babich suggests, not entirely accidental: 'No one less than Don Ihde,' Babich elaborates, 'arguably the preeminent American philosopher of technology, blocked the translation of Anders's work, incidentally ensuring that it would not be well known to this day.' See Babette Babich, '*O, Superman!* Or Being Toward Transhumanism: Martin Heidegger, Günther Anders, and Media Aesthetics', *Divinatio* 36 (2013), 42–99, 46.

55. The other three essays are (1) 'Die Welt als Phantom und Matrize: Philosophische Betrachtungen über Rundfunk und Fernsehen' (*die Antiquiertheit 1*, 97–211). A drastically abridged version of this uncanny vision ahead into a world that was about to witness the introduction of widely affordable television sets is available in English translation: 'The World as Phantom and Matrix', trans. Norbert Guterman, *Dissent* 3.1 (Winter 1956), 14–24. (2) 'Sein ohne Zeit: Zu Becketts Stück *En attendant Godot*' (*Antiquiertheit 1*, 213–32). A translation of this essay is available: 'Being without Time: On Beckett's Play *Waiting for Godot*', in *Samuel Beckett: A Collection of Critical Essays*, trans. Martin Esslin (Eaglewood Cliffs: Prentice-Hall, 1965), 140–51. (3) 'Über die Bombe und die Wurzeln unserer Apokalypse-Blindheit' [On the Bomb and the Roots of our Blindness toward the Apocalypse], an essay that introduces a way of thinking about nuclear weapons that Anders would extensively develop in its wake (*die Antiquiertheit 1*, 233–324). An abridged version is available in English as 'Reflections on the H-Bomb', trans. Norbert Guterman, *Dissent* 3.2 (Spring 1956): 146–55.

56. Anders, *Obsolescence 1*, 14–15.

Part I

ON PROMETHEAN SHAME

Preface to the Translation of 'On Promethean Shame'

In a 1979 interview, Günther Anders recounts how those émigré German authors who had 'cultivated their own style of writing' – he names Berthold Brecht, Thomas Mann, Alfred Döblin and tentatively adds himself to the list – had 'much greater difficulties than some others' to begin writing in English or to oversee translations of their own work.[1] As if to prove a point, Anders employs the eminently untranslatable word '*Eigensprachler*' to refer to this group of writers who were reluctant to leave the medium of their native tongue. An *Eigensprachler* is 'someone who has personalised language' and has expressly turned it into 'his own'. This might make it seem that Anders indulged in a 'private language', one that only he could understand. Nothing, however, could be further from the truth. The use of diary entries, the inclusion of poems, the persistent linguistic wit and wordplay, to name some of the traces of Anders's 'personalised' approach to German (and philosophical argumentation), have the rare trait of clarifying rather than obfuscating. The language itself is a key component of Anders's method of 'philosophical exaggeration', of creating 'overstated portrayals pointing in direction of the truth'.[2] Like the motifs of these at times absurd images themselves, the style in which they are linguistically presented is designed to reconfigure the manner in which the reader affectively and rationally relates to the subject matter depicted. Anders's writing is hence activist not only in content, but also in form, as the latter attempts to activate the reader's imagination as well as his or her critical and intellectual resistance. In line with this, sentences are at times incomplete or aborted, with the consequence that the reader has to bring them to their implied conclusion. The principal aim of my translation of 'On Promethean Shame' is to preserve the overall effect and the immediacy of Anders's exciting and eccentric use of German. As a rule I have chosen accuracy over elegance, not least because this gives an impression of Anders's idiosyncratic use of German. This has inevitably meant some compromises.

Anders's frequent use of the word *Mensch* (human being) presents the translator with a major challenge. In the service of readability, I translated this in the plural as 'humans' throughout, as I wanted to avoid using 'man' or 'mankind'. I adopted the same approach when the text makes a general claim by using a word in the singular ('the worker', for example). Translating this in the plural ('workers') was necessary, as German nouns are gendered and I sought to avoid – wherever possible – cumbersome formulations incorporating 'his and her' to signal that the text is not simply referring to male workers, but to workers in general, regardless of gender. Where a specific individual seems to be implied, I kept to the singular word form. 'On Promethean Shame' also makes use of several words for 'machine': *das Gerät, der Apparat, die Maschine*. These have slightly different nuances in German. Besides being a reference to the world of machines in general, *das Gerät* evokes technological objects that have become part of everyday life to such an extent that one might not think of these objects primarily as machines, but simply perceives them as things that one makes use of. Similarly, *der Apparat* is often a reference to a specific technological device, a radio set, for example. As Anders seems to use these terms interchangeably, I have opted to translate all three with the word 'machine' unless a specific reference seemed to be implied. A variation of vocabulary to avoid repetition might have left the impression that Anders's essay seeks to articulate and enforce a conceptual difference between the three terms, which does not appear to be the case. The one nuance worth mentioning is that the everyday sphere of household and personal use that *Gerät* and *Apparat* evoke is partially lost with the word 'machine', which makes the translation less subtle in this respect. As for syntax, this often had to be rearranged. Anders's texts have a tendency to be written in long sentences comprised of interlaced subclauses that gradually build up to a concluding point that resolves the linguistic tension and specifies the meaning. This approach is impractical and highly counter-intuitive in English, and certainly does not have a similarly cathartic effect. To preserve the clarity and drama of the German, I usually divided long sentences into shorter ones and opened the resulting sequence of sentences with the concluding point to invoke the image that is then given substance. I hope this way of proceeding with the translation has helped to retain the vivid reading experience of the German original.

I incorporated footnotes wherever I thought that the reader would benefit from further elaboration on the translation. I have also tried to retain Anders's reliance on Latin and Greek expressions and have provided translations of these in the footnotes. Wherever possible, I sought to preserve the formatting of the German original. I have replicated the many textual emphases (quotation marks, dashes and italicised words) that Anders deployed somewhat eccentrically. As in the German original, Anders's own extensive notes are given as endnotes. Anders concluded the introduction of *The Obsolescence of Human Beings 2* by quoting Max Weber's remark that 'the most important things, of course, are written in the notes'.[3] This comment can serve as an indication that the endnotes significantly add to the argument, often by elaborating on broader philosophical contexts.

I would like to thank Professor Chris Weedon for her patient and highly meticulous review of my translation and the extensive and invaluable editorial suggestions that emerged from this work. These have significantly enhanced the readability and accuracy of the translation. I would also like to take this opportunity to express my gratitude to Professor Rodney Livingstone for his kind advice on translating in general and for his many specific suggestions that have fed into the final version.

Christopher J. Müller

NOTES

1. Günther Anders in Elke Schubert (ed.), *Günther Anders Antwortet: Interviews & Erklärungen* (Berlin: Tiamat, 1987), 35. The 'others' Anders is here specifically referring to are his first wife Hannah Arendt and Herbert Marcuse. Anders did translate a few of his own texts into English, most notably perhaps are his 'Theses for the Atomic Age' (1960) and his long essay on 'The Pseudo-concreteness of Heidegger's Philosophy' (1949).
2. See Anders, 'On Promethean Shame', (ch1) 79* below.
3. Günther Anders, *Die Antiquiertheit des Menschen* 2, 14.

On Promethean Shame*

Günther Anders

If only I were one of you,
I'd no longer be ashamed;
I'd happily learn to enjoy,
my eternal return.

I'd run on an appointed course,
attuned to machines,
I'd do the same as yesterday,
tomorrow the same again.

No one would know, who I am,
no one would find out either,
who, in dark beginnings,
my parents were.

No one would know how I lay small in mother's lap,
after swimming like a fish,
that as a bloody lump of clay,
my life on earth began.

If only I'd have glowed in the oven,
and been made by things,
so as to spring amongst your ranks,
with an armour plated disposition.[1]

* The source text for the translation is: Günther Anders, 'Über Prometheische Scham', *die Antiquiertheit des Menschen 1: Über die Seele im Zeitalter der zweiten industriellen Revolution* (München: C.H. Beck, 2010), 21–95.

1

First encounter with Promethean shame – the Prometheus of today asks: who am I anyhow?

I begin with some diary entries from California.

March 11, 1942

Believe I have found the signs of an entirely new pudendum this morning; a form of shame that did not exist in the past. I will provisionally call it '*Promethean shame*' for myself. I understand this to mean the '*shame when confronted by the "humiliatingly" high quality of fabricated things (selbstgemachten Dinge).*'

Joined a tour together with T. of an exhibition of technology that has opened here. T. behaved in a most peculiar manner; so strangely, in fact, that I ended up only looking at him instead of the appliances on show. As soon as one of the highly complicated pieces started to work, he lowered his eyes and fell silent. Even more strikingly, he concealed his hands behind his back, as if he were ashamed to have brought these heavy, graceless and obsolete instruments into the company of machines working with such accuracy and refinement.

But this 'as if he were ashamed' is too tentative. The pattern of behaviour left me in no doubt. The things, which he recognised as exemplars, as superior and as representatives of a higher class of being, really did play the same part that persons of authority or of acknowledged 'higher' classes used to play for his ancestors. To stand in his bodily clumsiness and his corporeal imprecision under the gaze of such perfect devices was really unbearable for him. He was truly ashamed.

When I try to investigate this 'Promethean shame' further, then its basis, '*the basic flaw*' of the one who is feeling shame appears to be the nature of his own origins. *T. is ashamed about having naturally grown instead of having been made*. He is ashamed because he owes his existence to the blind and uncalculated, the highly archaic process of procreation and birth, which places him in stark contrast to the immaculate products, which are carefully designed through and through. His shame thus consists in his '*natum esse*,'* in his lowly birth. He regards this as 'low' (not unlike the chronicler of founders of religion) precisely *because* it was a birth. Once he is ashamed of his antiquated line of descent, the same naturally also holds for its faulty and inescapable result: *he is ashamed of himself.*[2]

---

* *natum esse*: having been born – Trans.

Incidentally, T. kept silent throughout the tour. He rediscovered his voice only long after we had left the exhibition. This seems to corroborate my shame hypothesis further, because any form of self-expression would have contradicted shame: if shame shows itself at all, then it does so precisely through self-concealment.

March 13

Reflections on T.'s behaviour:
Promethean defiance is the refusal to owe anything, including oneself, to anyone else. Promethean pride consists of seeing everything, including oneself, as one's own achievement. Some vestiges of this stance so typical of the self-made man of the nineteenth century are certainly still alive today. But I doubt they are still characteristic of us. It seems that feelings and attitudes of a different kind have taken their place; attitudes resulting from the peculiar fate of Prometheanism.

Prometheanism has experienced a truly dialectical reversal. Prometheus's triumph has in a sense been all too overwhelming. It was so triumphant in fact, that Prometheus is beginning to lose the pride which was a matter of course in the previous century when confronted with his own achievements. A sense of abject inferiority and self-contempt is replacing this. 'Who *am* I anyhow?' the Prometheus of today asks himself while playing the jester at the court of his own machines. 'Who am *I* anyhow?'

It is against this changed background that one needs to consider the desire of today's human to become a self-made man, to become a product: *Humans do not want to make themselves because they can no longer tolerate anything that was not made by them, but because they too no longer want to be someone who is not made. They feel indignant not because they were made by others (God, gods or nature), but because they were not made at all and as such they are inferior to all their fabricated things.*

March 14

What's at stake here is clearly a variant of a classic error: namely a '*confusion of creator and creation*'. This expression has its origin in the *Confessiones*. Augustine uses it to describe *the* religious error *par excellence*: the worship of a *creatum, that is*, of something that was created (the world or a worldly artefact), or of the pictorial representation of a god in the guise of such a *creatum*, instead of reserving this worship for the *creator* who alone is deserving of it. The equivalence of the two errors is clear: when feeling 'Promethean shame' humans also prefer what was made over the maker. Here too, the creation is accorded a higher state of being. With this, however, the analogy comes to an end, because the role that humans play in 'Promethean shame' has completely changed: whereas for Augustine humans belong *eo ipso* amongst the animal kingdom, they now appear only in their *homo faber* quality: as beings, therefore, who fabricate things. 'Confusion of creator with creation' in this instance means that the honour humans extend to their things is actually due to themselves, and to themselves alone. The substitution is here not proof of undue pride, but of self-degradation. Humanity denies itself the credit it deserves. T. attests to this self-degradation with his behaviour.

A legend from Molussia* recounts how Bamba, the Molussian god of creation, immediately sought refuge in invisibility after he had formed the Molussian Mountains. It is said that 'from that day onwards he remained invisible because he was ashamed not to be one of the gleaming peaks he had created'. T.'s behaviour matches that of this Molussian god.

March 15

T. certainly also wants to 'exist' ('*Da*' – '*sein*').† But to exist as a 'natural son' rather than as a legitimate product; to exist as someone who was fathered rather than fabricated, to be a human instead of being equal to other, precisely functioning, convertible and reproducible machinery – that's what pains him, that's what embarrasses him. Probably he even deems all this to be an original sin (even though he does not make use of this term and does not express his disillusionment in words at all). In short and in the words of an industrial song from Molussia:

Each day out of machines arise
ever more beautiful machines.
Only we remain malformed,
only we're born obsolete.

Too soon from obscure ground,
preformed and cast-out,
at this late hour we humans stand
imprecisely in this world.

Alas in view of machine-precision,
a head held high becomes us not.
Only things are granted confidence,
only machines are granted pride.[3]

---

* Molussia is a fictitious land invented by Anders (see the first endnote for further details). – Trans.
† Anders is here playing on the German '*Dasein*' (existence), which literally means 'to be here'. This has added (implied) significance because *Dasein* is the term Martin Heidegger deploys in *Being and Time* to suggest that humans exist in the sway of an experience of what it means to 'be' – Trans.

2

Three reservations are discussed: Promethean shame is absurd. It is invisible. It is trivial.

After I had read out these diary entries, the claim that a new – hitherto unknown – variety of shame exists or is at least emerging was met with the following three objections:

First Objection: 'We humans have made these machines ourselves. Our natural and legitimate attitude toward them is therefore pride. The assumption that *homo faber* must blush to see what he himself has fabricated, or that the producer would feel shame of the sight of his or her products, is absurd.'

Riposte: This challenge is rhetorical. The unifying words 'we' and 'our' express a purely verbal solidarity. That 'we' are proud isn't true. Who is this 'we'?

If anyone had the right to employ such a 'we', then it would be the minority of researchers, inventors and experts who truly master these mysterious realms. *We*, however, the remaining 99 per cent of today's population did not 'make' the machines (the cybernetic ones, for example). They are not felt as 'our' achievements, but strike us rather as strange and disconcerting. One's own products are never disconcerting. Even if we had been involved in the making of these things, we would not feel like their proud creators. Production processes are today made up of so many single and separate steps that the workers who actually build the world of products and machines cannot see these fabrications as the fruits of 'their own' labour. There is no longer any occasion for pride – be it for a single end-product or the world of machines and products as a whole – because no end-product shows any sign that it was invested with the specific qualities and achievements of individual workers. One can only be proud of achievements that bear traces of individuality that make processes of identification possible. For the workers, the world of machines is no more an object of pride than it is 'their' property to own. This, of course, is all the more true for all those who are not involved in production processes at all.

No, for the majority of people today machines do not primarily exist as human products, and certainly not as a testimony to some personal Promethean glory. This is the case despite the tremendous increase and ever more widely shared competence in the use of technology, and notwithstanding the common knowledge that products do not grow on trees. Products, rather, are simply 'there'. We encounter them primarily as necessary, desirable, superfluous, affordable or unaffordable consumer goods which become 'mine' only after I have bought them. As such, they are more likely to be proof of one's own insufficiency than evidence of one's own power: in a highly industrialised country the excess of displayed yet unobtainable products is simply overwhelming: the shopping mall is the permanent exhibition of what one *does not* have.

As for the man who is confronted with a working computing machine for the first time, self-aggrandisement and pride are even more alien to him. An observer who erupts with the exclamation 'my goodness, aren't we great guys, to be capable of this!' when encountering such a machine is a clown, a figment of imagination. Quite the contrary! He rather murmurs with a shake of his head: 'My god, it's incredible what *it* – the machine – can do!' At the same time, he feels highly ill at ease in his creaturely skin: for the machine half gives him the creeps and half puts him to shame.

Second Objection: 'I have never been able to observe the onset of this Promethean shame.'

Riposte: Entirely possible. Only rarely can it be caught sight of in action. In most cases one must deduce it indirectly from modes of behaviour (provided that the one shamed by products does not detect this feeling himself). This has the following reasons that are related to the nature of 'shame'.

I. The best known varieties of shame (e.g., the shame associated with the genital area) become acute in the interaction of humans and become visible – as barriers to communication – in encounters between human beings. 'Promethean shame', however, is a form of shame that is felt in response to a thing. Because no human partner is present to make us feel shame, there is likewise for the most part no human partner to witness our shame.
II. Shame does not 'appear' at all when it sets in.[4] When shame 'appears', it does not 'step onto the stage of visibility', it 'takes a step back'. Humans who are ashamed try to conceal their flaw and themselves. They are, however, unable to fully realise this desire to 'sink into the ground and disappear completely'. From this wavering shame two peculiar, if you like, 'dialectical' consequences ensue which explain the 'invisibility of shame'.

First consequence: Because the one who is ashamed remains visible, and because shame draws attention to its presence, a second form of shame immediately compounds the first: *the shame of being ashamed.* Shame accumulates ('iterates') automatically. It is nourished, so to speak, by its own flame, which becomes hotter and hotter the longer it burns.

Second consequence: In order to put an end to this self-accumulation of shame, which becomes increasingly unbearable with every passing moment, the one affected resorts to a trick: *instead of concealing his flaw and himself, he conceals his shame*, and even his gestures of concealment. He adopts an attitude that is diametrically opposed to shame, brazenness, for example, or one that declares 'I couldn't care less'. In order to conceal his shame, he thus practically rips the shirt from his own chest. With this motion he often not only wants to deceive the one before whom he is ashamed, but also himself. It is not only in individual (erotic) situations that this mechanism can be observed: in America's love life of today there is no more powerful motif than the shame of still being considered puritanically coy. Result: the bikini. In other words, in order to overcome his or her urge for self-concealment, the shamed person decides to return to the level of the most normal visibleness. 'Anyone who does not conceal him or herself', or so this person calculates, 'anyone who remains conspicuous is not suspected of shame (and thus also not suspected of having a flaw).' He or she hence behaves like someone who is wearing a torn item of clothing but who remains confident enough to treat the tear as non-existent. If shame remains imperceptible and invisible, then this is so because it is '*concealed through visibility*'.

Third objection: It is possible that T. was ashamed. In this case, however, this 'Promethean shame' is not an exciting new phenomenon, but rather the sign of a well-known state of affairs: it is merely a symptom of the 'reification and objectification of the human', which has been discussed ad nauseam.

Riposte: No. It is a sign of much more. What T. deems to be the occasion of his shame is precisely not that he has been 'reified', but the other way round. He is ashamed because he has *not* been reified and because he is not a thing. That's also why the term 'thing-shame', which I had initially considered, is out of the question. *A new, second level in the history of the reification of human beings* has been reached with this response of feeling *shame at not being a thing*. Humans now acknowledge the superiority of things, bring themselves into line with them, and *welcome their own reification*. This also means that non-reification is condemned as a shortcoming. Or – if one deems that this stage has already been reached – *a third level* must be considered: one in which the response of the one who is reified (approval or rejection of reification) has already become second nature and has become so immediate that it is released not as a deliberate judgement but as a feeling. This is certainly something entirely new – and T. has clearly reached this stage: intimidated by the superiority and supremacy of products, he has already *deserted to join their camp*.[5] Not only does he now share their point of view, not only has he fallen in line with the standards they set, his *feelings* have also become theirs: he despises himself in the same way that things would despise him if they could. And when he feels shame, he does so precisely because of this contempt, which he deems well-deserved. Such self-degradation in front of fabricated things has not recurred since the end of idolatry.

3

An Example of the Reification-of-Self: Make-Up

Two expressions used in the above ripostes require clarification: 'endorsement of one's own reification' and 'desertion to the camp of machines'.

a) An example for the endorsement of reification: '*make-up*' as a reification of self.

For girls, it is out of the question to go out in public without make-up. This does not only mean, however, that they are embarrassed to show themselves in a scruffy or unadorned state, as their mothers and grandmothers were before them. No, the pivotal point here is *when*, that is, in what state of adornment, they feel attractive. *When* are they deemed to be 'well turned-out'? *When* do they believe that they have no reason to be embarrassed? Answer: once they have transformed themselves into *things*. Once they have changed into crafted objects, into manufactured goods – as far as the raw material of their limbs and faces allows this to happen. To show oneself with 'naked fingernails' is 'impossible': fingernails only become fit for society, fit for the office, yes even fit for the kitchen, once they 'match' the appliances with which these fingers have dealings; only when they show the same dead, polished thing-'finish' as things and when they come across as manufactured. In short, the fingernails have to *disown their organic past-life*. The same standard also holds for hair, legs, facial expression and indeed the body as a whole (although, in this last instance nature is a bit recalcitrant). *Today it is the unmodified body that is 'naked', and not the unclothed one*: the body, then, that does not sport any thing-like qualities and does not bear signs of reification. One is ashamed of this body that is 'naked' in a new sense even when it is dressed; much more so than one is ashamed of a body that is 'naked' in the conventional sense, as long as the body on show has been reified in a pleasing manner. Every bathing beach demonstrates this, not just fashionable ones. To vary a famous word by Nietzsche: The body is something 'that must be overcome'. Or rather: It *already has been* 'overcome'.[6]

4

Examples of the 'desertion' of humans to the camp of the machines – the first calamity of humankind: the human body is 'stubborn and rigid' – the 'stubbornness' of their body turns humans into the saboteurs of their own achievements.

b) Let us now turn to the expression: 'humans are deserting to the camp of their machines'.

What was meant by this metaphor?

An American air force instructor teaches his cadets that humans, in their natural state, are '*faulty constructions*' when measured against the tasks they face.[7]

Regardless of whether the instructor was joking or serious, a better attestation of the 'desertion' in question could not be thought up. Humans can only be considered constructions, 'faulty' ones even, once they are classed as machines. Only when the category 'machine' has become universally applicable and is deemed all-inclusive, can that which was not constructed be reinterpreted as a 'faulty construct'.

It cannot be denied that humans are inferior to their technological appliances in terms of strength, speed, and precision, in the same way as human thinking power fares badly in comparison to the thinking power of 'computing machines'. From his point of view (which he has borrowed from machines) the instructor is thus right. And this even more so, because he no longer even has this type of inability to compete in mind. He would probably already deem it embarrassing to occupy himself with this comparison. No, he is already making a much more modern comparison: he is not thinking of *humans as machines existing alongside other machines*, but of *humans as machines existing for the sake of machines*. He is considering humans as working-parts of machinery that is already built, or as parts of already specified blueprints for future technology. As such he calls humans 'faulty' – which means 'fitting badly', 'inappropriate' and 'not measuring up', that is, not made to suitable standards. This being 'not made to suitable standards' could still mean that humans are merely unformed, unshaped raw material. But they are not even this. For humans *are* preformed, they *are* pressed into a shape, they *have* a form: just the wrong one. Not only are humans not made to specification, but, because only 'matter' can be reworked, they cannot be made to fit the specification. Unless already shaped humans are demoted to the status of mere material, are regarded as 'scrap metal', are treated as such, and are melted down. With this image, however, we are already getting ahead of ourselves.

From a machine point of view, the inability to be made to measure (*anmessen*) naturally appears to be a human 'pretention' (*Anmaßung*), yes, even a scandalous resistance and revolt. And in the mind of those who have turned the measure of machines into their own benchmark, a disgraceful failure.[8] Even if it was just meant as a clever remark, behind the instructor's phrase 'humans are faulty constructions' lays the indignation of machines and humanity's shame. 'After all' (one would probably be allowed to let the instructor say) 'we cannot expect our designs and machines to undergo alterations just because *we*, as human components with limiting requirements, are made so unreliably.'

Nothing would now seem more obvious than the following conciliatory response: 'regardless how great the superiority of machines is in strength, speed and precision, they are and remain mere rigid and definitive objects. Whereas we humans – yes, and this constitutes our dignity – are alive, malleable, adaptable, elastic and hence free.' But precisely this would be contested by the air force instructor. 'It is the other way round!' he would exclaim. 'When I call us humans "faulty constructions," I do so precisely because we are rigid and unfree in comparison to things!'

What does he mean? How did he arrive at this conclusion?

Let's forget for a moment the distinction between a dead object and a living human being that is the foil for our pride.

Our world of fabricated products does not define itself as the sum of single, finalised pieces, but rather as a process: the daily new production of things which are daily new. The world of products does hence not 'define' itself at all. It is indefinite, open, malleable, daily intent on reforming itself, daily adaptable to new situations, on the verge of new tasks every day. Reshaped continually through trial and error, the world of products presents itself every day anew and different.

And we? And our body?

No sign of daily change. Our body is today still yesterday's body, still the body of our parents, still the body of our ancestors. The body of the human who builds rockets can hardly be distinguished from the body of the troglodyte. It is morphologically constant; morally speaking: unfree, stubborn and unaccommodating. Seen from the perspective of machines, it is conservative, anti-progressive, obsolete and un-changeable: a dead-weight in the rise of machines. In brief: *we must invert the relation between freedom and unfreedom. Things are free and human beings are unfree.*[9]

The air force instructor had this and nothing else in mind when calling humans 'faulty constructions'. His figure of speech was much more than a witty remark. For, if we assume that the future tasks that will fall to technology are close to his heart – and we have no reason to doubt this – then we can see his words voiced by a grave concern: *the anxiety that our body*, regardless of whether the rift between it and our products broadens or even if it were merely to remain constant, will *threaten all new projects* that new machines show to be necessary; that our body will end-up wrecking one project after another.

'In principle', or so the instructor could argue, 'if it only depended on machines, we could soar to such altitudes – but this "in principle" is redundant – because *we* couldn't survive it anyhow, because our body would go on strike, and so we can't do it. "In principle," we could travel to this or that planet – but *we* are prevented from doing so by our own limitations – *we* cannot survive the stress test.' If the instructor had heard anything of Greek mythology he might conclude in this or a similar manner: 'The Icarus of today falls from the skies not because the wax holding his wings together fails, but because he himself, Icarus, is a failure. If he could only jettison himself as ballast his wings could conquer the skies.'

Not the worst description of humanity today, and the instructor would certainly subscribe to this, would be the following: '*Humans are saboteurs of their own achievements.*' For sure, they are not saboteurs in the sense that they would voluntarily harm their machines – in fact, despite all the weapons of destruction in existence no such idea has ever crossed their minds. Humans have never been more selfless towards any other entity than towards their machines. No, they are saboteurs, because they, the 'living ones', are rigid and 'unfree'. 'Dead things', on the other hand, are dynamic and 'free'. Humans are saboteurs because as natural products, as born bodies, they are too emphatically defined to keep up with the daily changing world of machines; a world, which makes a mockery of all self-determination. Deeply concerned, ashamed and with a guilty conscience, the human of today thus observes his highly talented children, whose future he fears to ruin because he can neither send them on their career-path without support, nor accompany them. And it often happens that he, the lame father, endows his fledgling children with servants that are designed to replace him, because he is bedridden with his earthliness and is unable to be part of their stellar rise and must stay at home. As a result, crews of escorting personnel and supervisors in the form of machines are assembled (e.g., the monitoring instruments he builds into his space rockets and artificial satellites). Or, to continue with a different image: the lame father sees himself forced to give them a 'boomerang-like' post to carry on their journey, mail made up of questionnaires overflowing with question marks, and furnished with the notice 'return to sender'. Mail that is tasked to return to earth completed to tell the story of what he himself would have liked to have learned along the way, if only he had been able to take part in the journey.[10]

The greater the misery of humans who produce goods becomes, the less they are a match for their own creations, the more relentlessly and tirelessly, the more voraciously and panicked they multiply the civil service that their machines, their sub-machines and their sub-sub-machines form into. By doing so, humans are naturally multiplying their own misery further: the larger, more intricate and complicated the man-made bureaucracy of machines becomes, the more futile are their attempts to keep up with it. It may hence be right to conclude that human misery leads to an accumulation of machinery, which in turn leads to an accumulation of human misery. Happy days, when the pastoral idyll of hydra was still seen as a frightful myth!

5

The human of today tries to escape this calamity by the means of physiological synchronisation with machines, by the means of 'human engineering' – the extreme perversion of supply and demand.

For the man of today it is of course out of the question to accept the stubbornness and inadaptability of the human body and to settle once and for all for human inferiority and retardation. He must, therefore, do something. His dream is naturally to become equal to his deities, the machines; or put more accurately: to be one and entirely part of them. Ὁμολογουμένως ζῆν.* As a Molussian hymn proclaims:

If only we could succeed,
to shed our burden and stand
like machine shafts
neatly slotted into others,

as prostheses with prostheses
most intimately bound,
then our defect would be overcome,
our shame would be unknown –

if this could still be granted us,
if this favour still could be bestowed,
no sacrifice in this world
would be for us 'too much to make'.[11]

Yet, as already said, this Molussian image of blissful '*co-substantiality with technological instruments*' remains wishful thinking. For humans, true integration into machinery is no more possible than real competition amongst equals. Machines are once and for all ontologically superior to beings who are born. This does not mean, however, that humans could wash their hands in their creaturely innocence, and let the matter rest, or that they would even want to do so. No religion has ever accepted the argument that the human is a mere creature and not a god as a *carte blanche* to be morally indolent. Neither the industrial religion of today nor its followers would accept the argument that 'humans are not products but merely creatures' as a valid excuse for being idly content with their own creaturely inadequacy. It is expected of humans that they prove their 'piety toward things' by at least attempting

* Ὁμολογουμένως ζῆν: to live as one with the gods – Trans.

self-reformation and an '*imitatio instrumentorum*'.* The minimum requirement is that the man of today makes the token effort to 'better' himself to such a degree that the 'sabotage' he practises *nolens volens*† (as a result of his 'original sin' of being born) is reduced to the smallest conceivable amount.

This is being attempted in the form of certain *metamorphoses of the self* called '*human engineering*', that is, 'engineering work on human beings'. We will describe a few examples in more detail. These attempted metamorphoses invariably begin as follows: the human engineer exposes his body to extreme and artificially created '*situations that test his corporeal limits*'.[12] To be able to study corporeal reactions, conditions are created that are *just about* bearable. Needless to say, these reactions are not examined in order to find out what the human physis is. For the objective is not to determine how this physis has 'naturally' delimited itself in the form of a body, and to then trace these limits cartographically. (In any case, a body subjected to unnatural conditions would hardly disclose its own natural limits. What is true for modern Physics hence also holds for our current example: the object of experimentation is modified by the parameters and mode of observation the experiment entails.) No, it is the other way round: the aim is to see if the human physique shows weaknesses and to catch such '*soft parts*' *in flagrante delicto*. I call parts 'soft' that can still be modelled and shaped, because they are currently still undefined, amorphous, changeable and ambiguous. Parts and areas, therefore, that might still allow the body to be adapted to the demands of machines.

The 'human engineer' does not want to know what the human physis *is*. He rather wants to find the threshold within which it *could* 'just about' *still be* (without biting the dust during the stress test). *He does not want to know how human physis measures up in its organically grown form*; but *which* alien *demands it could still 'measure up to'. Not what is fixed and certain about human physis is of interest*, but *which of its limits are not fixed and can still just about be altered.* He is hence only interested in artificially induced corporeal limit situations, because he aims to surpass these limits. If he has found a situation that is 'just about bearable' and, through dressage of the self, has managed to become accustomed to it, then he fraudulently moves the limits to a new threshold. A new 'limit situation' is staked out that he must become accustomed to through training his body.

Like a pioneer he endeavours to push his limits further forward, ever further away from himself, 'transcending' himself more and more.[13] Although he does not step across into the supra-natural, he does enter the no-longer-natural. By leaving his innate limitations behind, he enters the realm of the hybrid and artificial. In brief, *the aim of the experiments is to make the human*

* *imitatio instrumentorum*: imitation of instruments – Trans.

† *nolens volens*: like it or not – Trans.

physis undergo a metamorphosis. With the exception of magic and medicine, this physis has always been regarded as 'fate'. The aim is hence, to *strip physis of its fate, of all that is fatal.* This also involves the removal of all that is 'embarrassing' and shameful (as the word 'fatal' hints at).* Who knows whether the eagerness with which these great-grandchildren of the Puritans dedicate themselves to this masochistic transformation of the body is not in part explained by the secret current of residual energy generated by the puritan hatred of the body, energy that they do not recognise as such and which cannot be used otherwise in today's world.

The 'theory of the body' has thus been replaced by a specific '*praxis*', by a '*physiotechnology*' (if this analogy to 'psychotechnology' is permitted). Not, however, a 'physiotechnology' of the kind we already know from medicine, but rather a new revolutionary technology intent on toppling and abolishing the presiding 'regime' of physis as such. The aim of this technology is to shape from the 'existing state' of the body radically new ones. Following a famous pattern, its maxim could be described thus: '*It is not enough to interpret the body in various ways, one must also change it.*' And this needs to be done again everyday and each time to fit the specifications of the machine in question.

If a machine requires a body capable of operating in icy conditions, then one exposes one's body to the freezing cold in an icing-up installation which is specifically built so one can train in these icy conditions and transform oneself into a being that is their match.

If another machine needs endurance of abnormally low-pressure, then one builds low-pressure chambers in order to force a metamorphosis into a low-pressure-being.

If a third requires resilience to centrifugal forces, then one builds a centrifuge that pulls one's innermost parts outward until one has let oneself be transformed into a being that can live in such ex-centricity.

The first step in this process invariably involves the determination of the pre-critical point, the environment of the 'just about bearable'. The second step is to train and exercise in the said environment and the third to push the threshold further '*behind the back of physis*' once the exercise has succeeded. Human ingenuity really knows no limits here. In point of fact, there are no catastrophic conditions that could not be fabricated, or rather modelled in true to life (this actually means true to the machine) training- and martyr-simulations. No ingenuity or luxury is spared in the construction of these models, for the human engineer has staked everything on them. With their help he hopes to be able to pass the high school diploma for machines (*Geräte-Abiturium*) and in doing so to be absolved of his shame by the machines he admires.

---

* In German, the word *fatal* also carries the meaning 'really embarrassing' – Trans.

What the body is to become is in each case determined by what the machine demands. The relationship between '*supply and demand*' has here experienced a peculiar perversion, one that far exceeds the conventional perversion of this relationship that is already normal today.

The 'conventional perversion' of supply and demand, as I understand it, has two aspects:

(1) *The inversion of the chronological sequence*: The fact that it is not usually demand that leads to supply, but the other way round – supply leads to demand. (2) The fact that *demand is also produced* as a '*secondary product*' which is essential for the sale of the actual product being supplied.[14] One thing, however, is still left unaltered in this normal perversion, or so it seems at least, namely the division of roles between demand and supply: *humans are here still being* treated as '*subjects of desire and demand*', even if as consumers they are terrorised or flattered into their needs.

The destruction of this semblance is the new element that perverts the relationship between supply and demand further. The *machine* now openly and shamelessly presents itself as the '*subject of desire and demand*'. It demands to be offered what it needs and expects that humans exert themselves to come up with ever better offers (because humans in their present state are not acceptable to machines). Humans are thus required to supply everything the machine needs to function as well as it 'could' function. This 'could' is not merely a subjunctive clause suggesting that this 'could' does not commit us to anything. No, this 'could' describes what is 'due' to technology, and the meaning of the word 'due' already shimmers across into being an obligation.

Into being an obligation. This means: moral demands are now also transferred from the human to the machine. What is 'due' now also becomes what one 'ought to be'. The maxim 'become who you are' is acknowledged to be the valid maxim of machines. Human purpose in life is now limited to guaranteeing the success of this machine-maxim through the offer, provision, and preparation of his or her own body.

Only one generation ago there were many phrases (in pedagogy and the branches of psychology concerned with aptitude) that sought to express the sentiment 'give the capable a free hand'. Such phrases convey the view that the 'potential and talents' of humans are sacrosanct in themselves. They imply that it is a moral duty to develop and make use of these talents and that it is immoral to neglect or suppress them. What was applied to humans yesterday is now reserved for the machine: its 'potentials' are nurtured. It is now a duty to 'give it a free hand', the stifling of its 'potential' has become an immoral deed. *Machines are the 'gifted talents' of today.* Even the inherent 'potential' of the H-bomb, the wunderkind amongst machines, is held to be sacrosanct. As diverse as the excuses for the existence of the bomb may be, a good part of the vehemence with which protests against its existence are met, can be explained by indignation at the fact that there are humans who take the liberty to oppose the maxim 'become who you are', who want to nip the 'organic development' of the machine's potential in the bud and block its 'free development'. Today, even the most dreadful machine can be successfully justified and defended if its critics are suspected of being Luddites. And since nothing is easier, this always succeeds.

6

Human Engineering: Initiation Rites of the Robotic Age – 'Dehumanisation' does not alarm the dehumanised, for it does not belong to their area of responsibility.

The reader who comes across the details of these human engineering experiments for the first time is plagued by an irritating feeling of déjà vu, or rather by a feeling of *déjà lu*.* The irritation lasts until he or she suddenly realises what it so strikingly reminds him or her of: namely, ethnological accounts of *the rites of passage* which tended to precede the admission of adolescents into the community of adults in the life of a clan. This association is certainly not unwarranted. For, what the human test subject hopes to achieve by subjecting their body voluntarily to torture-like experiments is really a kind of investiture. The test subject hopes to be admitted into the community of 'adults' and to be ranked amongst those who 'count'. Only it is machines that now count as 'grown-up'. *The experiments that make up 'human engineering' are* in fact *the initiation rites of the robotic age.* The human guinea pigs are candidates and eventually neophytes who are proud to have left their 'childhood' behind. They are proud to have undergone 'the education of humankind' that is demanded today.

However, if 'machines' are considered to be 'adults', then 'to move on from one's childhood' and 'the education of humankind' means 'to leave the human condition behind and stop being human'. For the moralist, who cannot let the idea of the human be swept aside, the trajectory of this development is the total catastrophe. What humans are hoping to achieve with their experiments is really the *climax of all possible dehumanisation.* I say 'climax', because for a century the most extreme image of dehumanisation, degradation and deprivation of liberty conceivable was that '*workers are being exploited by their working conditions*'. In the light of what has now actually occurred, this image has become the epitome of naivety and harmlessness.

* déjà vu […] *déjà lu*: already seen […] already read – Trans.

That in the context of human engineering the *passivum* 'being exploited by' is still aligned to shocking images, to images of suffering, unfreedom and misery is really out of the question. *What is deemed misery today* is, inversely, the limitedness of human 'usefulness and application', the *worrying possibility* that there might be *definite boundaries* to human passivity and usefulness, in short, *to human unfreedom.* The expression 'climax of dehumanisation' is certainly not an exaggeration: humans today do not shy away from any amount of blood, sweat and tears and use all their spontaneity and ingenuity to extinguish the spontaneity and human dimension of their performance and achievements. They have mobilised everything so they can implement their own reification and become passive, for they hope to explode the boundary of unfreedom that threatens them and finally obtain the *summum bonum** of total usefulness.

It is part of the nature of machines, that they conduct one single, specialised task that they are fully consumed by the performance of this task, and find in it the sole justification of their existence. Should machines display additional features beyond their intrinsic purpose – for example the heat generation of the light bulb – then these are merely 'put up with', if these attributes cannot be exploited further or eliminated. This aspect of machines now also defines the entity that is to be produced through '*human engineering*'. This being is to obtain one specific ability, for example, the capability to execute this or that manoeuvre at a certain altitude, at such and such a speed in specific low pressure conditions, within a certain response time. No matter how 'superhuman' such ability may be in comparison with what the body is actually capable of, the desired outcome is something *subhuman.* The 'actual human being' (if it cannot be eliminated completely) here merely becomes an unavoidable appendix to the machine, for the desired outcome of human engineering is a machine-function, a proficiency in an isolated 'skill'.

Of course, those who are part of 'human engineering' see nothing morally questionable in their activity: it is neither deemed 'outlandish', nor 'impossible', nor even dehumanising.

What strikes them as 'outlandish' is merely the fact that as humans they form such 'impossible' exceptions in a world that evidently belongs to machines and devices. And it is precisely this fact that makes humans conceive of themselves as 'hopeless', just as an inept machine is called 'hopeless'. As for the charge that 'human engineering' is dehumanising, this is completely incomprehensible to those who are involved in it. After all, to be like a machine is what they desire, and the task they have set themselves. For them, the robots clanking about in 'cartoons' disguised as puppet-like humans

* *summum bonum*: the highest good – Trans.

for fun are not undignified or terrifying figures, they embody their dreams and their duty to conform and synchronise.[15] Why the satisfaction of their yearning and the pious devotion towards things, why such pleasant and honest activities might be considered debasing, is destined to remain inexplicable for them. If there is anything at all that is bothering them about 'dehumanisation', then at most it is the fact that now and then old codgers appear, who think of attaching the ugly epithet 'dehumanising' to their activity. At most this. Usually they do not even notice such old codgers. Those who are invested in human engineering are not disturbed by the question of *what* (apart from the special-skill they drill themselves in to the point of virtuosity) *should become of them, as human beings* (for, the codgers suggest, they once were 'human' and, therefore, they should not deny that there is still something human-like attached to them as an irremovable dead-weight). Human engineers are not unsettled by such questions; indeed the questions are usually not even noticed since they do not belong to the area of their specialised experimental tasks. If such questions are acknowledged, then only in order to say that such matters belong to a different area of responsibility, and that someone else will have to take care of them. For, human engineers would certainly feel that to occupy themselves with such matters would 'interfere with their own work', make them neglect their duties, and that this would be immoral and ultimately even *indiscreet*. If one translates the unspoken moral code of specialists into words, then 'indiscretion' means that one has more interests than one's position requires, that one has mistakenly brought along a surplus of knowledge and interests not required for the job one is about to begin.[16] In short: in response to the question what should become of him as a human being, the human engineer would answer: that's '*his* business' while pointing his thumb back over his shoulder in no specific direction, for he has no clue to which authority, person or department this '*his*' relates; yes, he doesn't even know whether such a department exists. In point of fact, the expression '*his business*', which once meant 'everyone has the right to become happy or unhappy in their own manner' today means the following: 'if it is someone else's business to have an opinion about me, then I will take special care not to involve myself in his business (i.e., in my business that has been placed in his hands).' If this way of putting it seems too pointed, the following analogous example should be considered: today's relationship between doctor and patient. Patients are often not only prevented from knowing about their own state of health – about their proximity to death, they even voluntarily take no notice of it. For, the patient's state of health is the doctor's business.

Our human engineer would hence not get at all agitated, if instead of asking the question of what will become of him as a human being, one were to shout the ready-made answer in his face: 'you are turning yourself into human dross, are making yourself into a mere appendix of your artificially acquired specialist skill'. Either he would not register this attack at all (for he has immunised himself against everything that does not belong to his area of responsibility) or he would retort: 'you are barking up the wrong tree!' At best he would just shrug his shoulders, for he feels utterly confident in his cause. When it is a question of providing machines with what belongs to machines, he is certain that no sacrifice and no self-sacrifice can be called 'a scandalously unwarranted human sacrifice'. There is no chance that he could become a suitable target for attempts at moral intimidation.

7

The attitude of the transformed Prometheus: 'hubristic humility'

Never before, not even in their most mystical practices or magical transformations of the self, did humans attempt an operation that constituted such a radical '*rejectio fati*'.* Compared to the aims of human engineering, the building of the Tower of Babel appears to be an undertaking as meek as a lamb: the limit humans transgressed was here merely one pertaining to their workmanship and to the permitted dimensions of human products. Their transgression was hence quite 'innocent' because criteria that would have helped them discover the maximum permitted height and conform to it had not been defined. Since humans can produce things at their own discretion, such criteria are not at their disposal, unless *humanity itself* is taken as a measure. That is, unless they define the critical point as the moment when humanity, '*smaller than itself*', can no longer keep up 'with itself', that is, the moment when human beings are no longer a match for what they have produced.[17] This moment is now.

This much, at least, is clear: if humans disregard this critical point and begin to also alter their body, then this marks a fundamentally new, a truly unprecedented and vile chapter.

This, however, should not be misunderstood. It is not the alteration as such that we deem 'vile'. Anyone who hears in our arguments a hidden, 'metaphysically conservative' voice would misunderstand their meaning.[18] Nothing is further from my mind than assuming the position of a 'metaphysical moralist' who regards the existent (as it is or as he imagines it to be) as 'good' and as the 'preordained order of things'. I am not an ethicist who puts the morality of humans into the framework of what 'is as it is and must so be'. Nor am I someone who infers what is allowed and forbidden from the given order of things. The chances of developing such a 'metaphysical

---

* *Rejectio fati*: rejection of fate – Trans.

morality' have been forfeited a long time ago. It would be a futile exercise to want to retrospectively find a common denominator for questions of what is just (*questiones iuris*) and questions of what is (done) (*questiones facti*),* including those pertaining to metaphysical facts. To level the charge that 'reason' was split into these two parts by philosophy, in particular by Kantian philosophy, which established this split once and for all, is ridiculous (since Paulsen, synthesis enthusiasts, who regard everything to be organised like a box of toy bricks have mounted this attack on philosophy time and again). Because of the natural sciences, the world has transformed itself for us into a thing that is 'beyond good and evil'. The result is that moral problems, and not only theoretical 'problems', but above all *our own 'moral' and 'immoral' deeds*, are now, whether we like it or not, *drifting without roots in the ocean of morally indifferent being*. They are drifting, so to speak, as a '*metaphysical bunch of cut flowers*' that concerns nothing and nobody apart from us humans, and about whose inconsequentiality we are not allowed to harbour even the slightest illusions.[19] I say 'without roots', because the opportunity that Kant saw in the fact that we could turn our non- or only partial-affinity to morality into the metaphysically positive fact of 'freedom' has also been lost. Today, the life of a moralist is certainly no fun.

Far be it from me, then, to regard the form (*eidos*) or morphological fixedness of existing 'species' (the 'human' species included) as 'sacrosanct' or 'good' just because they are as they are, or to retrospectively sanction them (or condemn their transformation as a 'cosmic prohibition'). All the more so, because nature itself is clearly open to mutation and does not seem to place any great value on the permanence of species, in brief, because the existence of eternally immutable species has not been established. Even *if* there were such permanence, even *if* it were proven that the various species have endured unaltered to this day, this persistence would still not be morally conclusive, precisely because questions of what is just (*questiones iuris*) are independent of questions of fact (*questiones facti*). Or put otherwise, what exists does not have to be.

No, the alteration of our body is not fundamentally new and vile because we are abandoning our 'morphological destiny' or transcending the assigned natural limitation of our capabilities. It is despicable, rather, because we are undergoing this transformation of the self for the sake of machines and because we employ these as blueprints and models for our own modification. We hence relinquish ourselves as our measure as humans and with this gesture we either restrict or give up our freedom. It is for this reason, that the expression '*hybrid*' does not seem a suitable designation for what we are

* This distinction is made by Immanuel Kant in *the Critique of Pure Reason* – Trans.

creating, no matter how adventurous our experiments and their aims may be. The connotations of breeding that the word 'hybrid' evokes might be justifiable, for we are creating new hybrid-beings through the cross-breeding of fabricators and products; but is this also true for the ordinary sense of the word 'hubristic', namely 'arrogance' and 'pride'?* After all, acts of self-denial are not hubristic. And '*fitting oneself to a pregiven measure*' (*Anmessung*) cannot be equated to '*presumptuous self-evaluation*' (*Anmaßung*).† These acts are rather symptoms of curtailment, yes even of self-humiliation.

Even so, the impression is not easily averted that this act of curtailing oneself in the attempt to 'measure up to a machine' is also presumptuous. It would, of course, be unreasonable to dispute that the attempt to turn us humans into machine-like beings is monstrous. Were a 'human engineer' to ask us what we now took him for, presumptuous or unassuming, hubristic or humble, we would not be able to give him a clear answer. For, the phenomena 'of hubristically overestimating oneself' and 'of making oneself into a hybrid by fitting oneself to a machine's measure' are phenomena that are like (Siamese) twins who cannot be torn apart. The 'human engineer' is in fact both, 'overconfident' *and* 'self-deprecating'; both, hubristic *and* humble. His attitude is one of '*arrogant self-degradation*' and '*hubristic humility*'.

Of course these phrases sound mystifying. It is, however, only the constellation of words that is baffling, not the actual idea that they formulate. Just how old this idea actually is, can be seen, for example, in the prohibition of suicide shared by many religions. This prohibition is not justified on the grounds that the suicide victim is aiming for a titanic or godlike existence – for to suggest this would be absurd – no, it is the other way round. The person attempting suicide limits his or her own significance in such an absolute manner that in the eyes of religion such a judgement is the sole prerogative of the highest authorities. What is condemned is hence '*arrogant self-degradation*'.

The choice of this example is not a coincidence, since the self-degradation at issue in our discussion of human engineering is 'hubristic' in a very similar way to suicide. It does not aim at total bodily annihilation, 'merely' the annihilation of the human *as* human. Yet this 'merely' is debatable, since it is more 'hubristic' than suicide. Why?

Because the hubristic and hybrid intervention in the human that the human engineer performs does not just 'make use of what already exists', but invents something new. What does this mean?

---

* As used by Anders, the German *hybrid* acts as an adjective that can denominate hubristic or hybrid qualities. I have used both words to underline this – Trans.

† Anders is playing on the words for measure and measuring (*Maß*, *messen*) that link the words '*Anmessen*' and '*Anmaßung*', which otherwise do not share a direct link in meaning – Trans.

Whereas the suicidal person limits himself or herself to taking an inevitable event (dying) out of the hands of higher powers to bring it about preemptively, the intervention the human engineer dares to undertake is a first, a *novum*: it would never happen to him 'by itself'. It is a wrong that he would never be expected to face and suffer from the powers that be. His intervention does not amount to the appropriation (and the 'requisition') of an existing injury, but is rather a pretentious production of a new one. To the injuries he must expect from above, to hardship, illness, ageing and death he now masochistically adds a further one: self-reification. This is how one might imagine a theological account reflecting on this in the year 2000: 'Because the daemon or Marcionist God* who condemned humans to a machine-like existence or turned them into machines did not exist, humanity invented one. Yes, humans were even presumptuous enough to cast themselves in the role of this additional deity. They assumed this role, however, only to damage themselves in ways that they could not suffer from any other gods. Humans turned themselves into masters only to find a new way to become slaves.'

It may be true that humans are doing all this in the blind hope, or at least on the pretext, that this self-reification will reduce or even abolish all other forms of human frailty and degradation. But what counts are not reasons, but, rather, what is actually done: in order to combat the limitation of their liberty, humans limit themselves in a new way.

We have usually associated the figure of *Prometheus* with 'hubris'. For the last one hundred and seventy-five years (from Goethe via Shelley and Ibsen to Sartre's *Mouches*) our fathers treated *Prometheus* allegorically, and until recently we did the same. If we ask ourselves whether this figure still has validity and allegorical significance for the representation of our contemporaries who conduct 'human engineering', we arrive at the following ambivalent answers.

Our contemporaries are certainly still 'Prometheans', but strangely perverted ones. They also sneer: 'I should honour you – for what?'† But with this response they are refusing to honour themselves.

They also have presumptuously self-aggrandising ideas of entitlement – but these are so aggrandising that they begin to feel inadequate themselves.

They also suffer lacerations – but not because Zeus punishes their high-flying ambitions, but because they chastise themselves on account of their own 'backwardness' and the 'shame of having been born'.

Our contemporaries would of course recognise themselves even less in these mythological parables than in our account of human engineering in the

* The followers of the early Christian theologian Marcion (AD 85–160) held the belief that because Jesus is the embodiment of love he could not possibly be the son of 'the cruel and wrathful' (i.e. Marcionist) god described in the Bible – Trans.

† By stealing fire, Prometheus hubristically refused to honour Zeus – Trans.

previous sections. For them, their conduct is utterly non-mythical, or at least something that they can accommodate within the framework of the 'myth of progress' (although they do not realise that this is a myth). They automatically see 'progress' in each new step they take, even when this step takes them into low-pressure or simulated frost chambers. They are reassured because things are moving forward and because they are proud of this novelty.

After this digression, let us return to our more narrowly defined theme.

We had established: When humans suffer from feelings of inferiority in the face of their *machines*, then they do so, primarily because they are forced to realise – in their attempt to measure up to their machines – that they are a 'poor quality' raw material when it comes to being moulded into this or that machine part. Instead of actually being raw material, humans are 'unblessed' morphologically because they are already pre-given. Moreover, every pre-given form turns out to be 'wrong' and 'already pressed into the wrong shape', because in each instance the actual form machines require varies. In 'human engineering' people attempt to 'melt down' this 'misshapen form' in order to retroactively source the material from which the required shape at any particular time can still be made.

The fact of 'having already been formed in the wrong shape' is the main defect that the human engineer seeks to address and the main reason for 'Promethean shame'. It is only the main source, however, for he has many deficiencies that 'generate shame'. It would be impossible to compile a list of all these shameful shortcomings because this would correspond to a list of the virtues of all existing and all potential machines. In the following we will therefore limit ourselves to a discussion of the defect which seems to be of most consequence after that of 'being misshapen': the fact that *human beings 'perish easily*'.

8

The second inferiority of human beings: They perish easily and are excluded from the process of industrial 're-incarnation'. The human 'malaise of being unique'.

Although humans are less malleable than their products, humans are also shorter lived, more mortal. It is impossible for humans to compete with the durability – yes even the 'immortality' – they can give to their products if they want to.

Of course, the things we make are not 'immortal' in the strict sense of the word. The durability and shelf life of our preserved fruits, our frozen scrambled-eggs, vinyl records or light bulbs is limited. In most cases, however, it is *we humans* who have allocated mortality to our products and have limited their life expectancy (in order to secure a steady or even increasing demand for them, for example). Only our own mortality is not of our making. It alone has not been calculated. Because of this, it is a cause for shame.

With some restrictions, the use of the word 'immortal' to describe our products, even the most obsolete ones, could be justified. This is because a new variety of immortality has come into existence: *industrial re-incarnation*, meaning *the existence of serial products.* Certainly, viewed in the singular, the life, utility and performance of each individual item produced (*this particular* screw, *this* washing-machine, *this* vinyl record, *this* light bulb) has a limited time-span. But is this also true if it was produced as part of a series? Does not the new light bulb that replaces the burnt-out one continue the latter's life? Does it not become the old bulb?[20] Does not each broken or lost item continue to exist in the model, make and blueprint according to which it was made? Cannot every item console itself with the hope of being there anew, once its sister piece has taken its place? Has it not become 'eternal' on account of its replaceability, that is, because of the technology that makes this possible? O death where is thy sting?*

As strange as this may sound, it is only the vocabulary deployed here that is strange. The phenomenon itself is neither uncommon nor new. Example:

In contrast to the great fire which destroyed the Library of Alexandria, no page was actually burnt when thousands of pages were reduced to ashes in the course of Hitler's book burnings in 1933, because every page had hundreds or thousands of identical siblings. As ignominious as the intentions of the arsonist were, and as ominous the gesture of his hand which betrayed the fact that it would soon commit more than paper to the flames – at this stage, his acts of annihilation were still totally farcical. In the midst of the jeering crowd dancing around the pyres of books, there also danced unseen a feather-light group of hecklers, which the flames could not touch. Even while being scattered to the winds, this group of the '*book prototypes*' proclaimed: 'Go ahead, burn our copies! Burn them! Us you cannot burn.' And today these supposedly incinerated books live again and do so in thousands upon thousands of copies.

The world we live in today is certainly no 'Kingdom of Ideas'. But it is surely undeniable that it is more '*Platonist*' than any human world has ever been before. This is precisely because our world is made up of things that are for the most part mass-produced to a specific norm. These products owe their existence to *ideas*: they have come into the world as imitations or imprints of prototype models, blueprints or master copies. *None of the things* produced in this way now makes, or can make, the arrogant claim of being 'itself'. None is *individual and singular in anything other than a numerical sense*, for

* Corinthians 15.55 – Trans.

the ideas they incorporate also provided the blueprint for their sister pieces. As long as items produced to an unchanged model are kept in stock, as long as a new individual item can step in for a damaged one (for this specific burnt out light bulb here, for example), then it is merely a question of money, whether one deems the thing one possesses mortal or immortal. *For the man who is flush with cash, each thing can reincarnate itself in a new one*. And this prospect of reincarnation only ends when the 'idea' of the thing also dies, that is, when its make is dropped in favour of another one. The thousand copies produced from the blueprint of that particular make, of course, then also gradually disappear.

Of course, this industrial-Platonism, this 'immortality through re-incarnation' does not belong to the products themselves. They owe this virtue to us and this needs no further elaboration. This admission, however, does not provide a counter-argument to what has been said so far. For the object of this investigation is precisely the fact that we feel inferior to our products, *despite* the fact that we produce them ourselves. Here too, what counts is only our own disadvantage: only the fact that we cannot share the virtues that we bestow on our products. In this case the disadvantage in question is that none of us has been granted the ability to exist in the form of multiple (simultaneous or successive) copies. In contrast to the light bulb or vinyl record, none of us has the chance to outlive himself or herself in a new copy. In short: we must continue to live our lifetime in obsolete singularity and uniqueness. For those who recognises the virtues of the machine-world as exemplary, this is a flaw and as such a reason for shame.

That's right, a reason for shame. I am not using this expression as a metaphor here any more than I was last time. To show that the emotion here really is 'shame', I will share a particular experience:

About ten years ago I visited someone in a Californian hospital who was terminally ill.

In response to my 'how are you?' he made a gesture that seemed not only to include his immediate fellow hospital patients, but humanity as a whole. He murmured: 'There is not a lot happening with us, anyone of us.' He met my query about what he meant with a shrug of his shoulders, as if to indicate that the answer was self-evident. Then he added a rhetorical question to counter mine: 'Well [...] can they preserve us?' The word 'they' referred to the doctors; 'preserves' are bottled fruits – he wanted to say 'can they perhaps turn us into preserves?'

I said 'No.'

'And,' he continued, 'spare men they haven't got either.'

'Spare men?' I asked not understanding.

He elaborated, 'Well don't we have spare things for everything?'

Now I understood. He used 'spare men' as an analogy with 'spare tyres' or 'spare bulbs'. Like another light bulb, so to speak, that one could screw into the fitting in his place once he had burnt out. What he meant to say was 'they do not keep spare men in stock for us'.

His last words, however, were: 'Isn't it a shame?'

He was thus suffering from a double feeling of inferiority: (1) that he could not be preserved like a fruit and (2) that he could not be replaced like a light bulb. For he was quite simply – and here the disgrace was undeniable – a categorically perishable individual piece.

It was not easy to console this man. Who on earth is mentally prepared to see a fellow human being suffer because he can neither be preserved like fruit nor replaced. The words of comfort I improvised were certainly inadequate. And of course his distress in shame and in death was in no way less than that of others in their final hours. It cannot be denied either that he also had a particular experience of mortality, yes, that he even used 'eternity' (*aeternum*) or at least 'being everlasting' (*semperitum*) as a foil for his own.* Only that it was no longer god, the starry heaven, the realm of ideas or humankind that now served as a foil (while I sat by his bed-side holding his hand, the considerable differences that no doubt exist between these terms seemed to dissolve). Instead he turned to the world of preserved peaches and branded products that have become immortal through reincarnation. This was thus the eternity for which he had respect, this the foil in front of which he was ashamed. In this state of shame, the light of his life went out. And because he could not ascend to the Mount Olympus of manufactured goods, but plunged into the Hades of raw material, he ended-up more destitute than all the fabricated things he had valued during his lifetime.

There is nothing more rebellious and cunning than us, the people of today, when we think we are being cheated out of even the merest shade of unfreedom. So of course we naturally do not submit without resistance to this *numerus clausus*† and to our exclusion from the class of reproducible branded goods. We are no more willing to put up with our 'faulty construction' than with our irreplaceability and existence as unique beings.

'Why,' someone could object, 'why do we still need this? Aren't we all what we eat and "what we do," and since millions of us already consume and do the same things, hasn't our indistinguishability – and with this our interchangeability – reached its peak? Are we not also only "individuals"

---

* *Aeternum* [...] *semperitum*: 'something eternal' [...] 'something that is always'. Anders is here playing on the nuances of these two Latin words meaning 'always'. The former evokes the eternal realm of the 'afterlife' whereas the latter has no such transcendental resonance – Trans.

† *numerus clausus*: 'the fixed number of places' on specific courses, which dictates competition for entry to university – Trans.

in a numerical sense? What reason could we still have to complain about a *numerus clausus*?'

Viewed from the perspective of institutions, the economy, the leisure industry, politics, and warfare – all of which make use of us as instruments of labour, consumers or victims of some other kind – this replaceability is already an undeniable fact. It is, however, not merely a bare fact, but one that is daily confirmed by public opinion and a state of affairs that is also expressly approved of by scientists. For even social psychologists and moralists have got nothing better to do than idealise 'adaptedness' and 'inconspicuousness', that is, to present everyone who still insists on a rudiment of selfhood or a trace of personal difference as a 'crank', that is, as a pathological weirdo.[21]

It cannot be denied that our transformation into reproducible mass products has already succeeded when seen from the perspective of these institutions (which make use of individuals). '*Spare men*' are always available (because everyone is merely a part and an actor in a labour process). However, this conclusion is only true from this institutional perspective. It is not shared by individuals themselves. Our already mentioned patient, for example, testifies to this. He did not complain about his replaceability; he grieved over his irreplaceability.

Each of us, every single individual, shares the fate of this patient. As replaceable as my contribution at work may be, as guaranteed the gapless continuation of my function, I myself, my identity as XY, cannot be continued by a substitute. Even if the 'I am me' which my substitute will utter corresponds verbatim with mine, it will now refer to another I, namely his. My own self will stay behind unreplaced and unreplacable.

Anyone who does not pay careful attention here can easily miss the crucial point. After all, it was the credo of all ideas of humanity that the singular as singular, that each individual as individual is irreplaceable (and 'irreplaceably valuable'). It could sound, therefore, as if we are here speaking about a basic characteristic of humanity or of what remains of the experience of being human.

The opposite is true. What is decisive here is *how* the individual confronts his or her own irreplaceability, *what* he or she feels it to be. What matters is that one deems one's singularity an undeserved handicap, an attribute that makes one recoil, even though one cannot deny being identical with it. In short, this individuality is experienced as a defect, as a *flaw* of which one is ashamed. This '*malaise of being singular and unique*' is decisive here.

If one gives this 'malaise of being unique' the trivial wording of a pop song – 'I'm unique, I won't come again' – then it is immediately clear that it is closely related to the fear of dying. And this makes sense: we said above that mass products have achieved a 'deathless' state due to their replaceability. If humans are excluded from this serial existence and replaceability, they are also excluded from their 'deathlessness'. The experience and knowledge that we humans are not mass-produced hence acts as a *Memento mori.*[22]

9

Humans attempt to escape this second calamity through 'Iconomania'

Now there is one phenomenon that offers overwhelming proof of how ubiquitous the feeling of disadvantage, this 'malaise of being unique', is today, namely, '*Iconomania*', the current addiction to images.

It is certainly indisputable that this addiction to images is a first in the history of humankind; as regards intensity and scale, this phenomenon has left all other current addictions way behind. Indeed, it is key to any theory of our age. Since only a new term can indicate that it is a concept of philosophical significance, we will coin the term '*Iconomania*'.

There is a utopian description of our world in words we have put into the mouth of an extra-terrestrial reporter. He states that the human-world appears to be, first and foremost, 'the vehicle and occasion of potential images'. 'At times,' he recounts, 'I was even under the impression that the purpose and function of the human-world is limited to this. At any rate, images play such a tremendously prominent role that only pure nothingness remains when I imagine their world emptied of its billions of images, of its photographs, films, TV-phantoms and posters. I have not met a single human being who did not voluntarily present several pictorial-duplicates of him or herself and their loved ones, no one who did not carry such images on his or her body or at least possessed the likes.'

One of the most important reasons for this hypertrophic production of images is that images allowed people to seize the opportunity to create 'spare-pieces' of themselves and, with this gesture, to belie their unbearable singularity. The excessive production of images is a large scale countermeasure against humanity's 'I am unique'. Whereas people are excluded from serial mass production, they transform themselves into '*reproduced products*' when they are photographed. *At least in effigies they can also attain a multiple, at times even thousand-fold, existence.* Even if a single human 'only' lives as the model for their own photographic-replica, 'he or she' somehow still exists in them.

It follows as a matter of course that *movie stars* (who are seen by more eyes than us common mortals) and are the most successful amongst us in assuming this multiple existence, become our role models. *We honour them with wreaths of laurels because they have made successful inroads into the sphere of serial mass products – a sphere we recognise as ontologically superior to our own human one.* We idolise them because they have in the most triumphant manner realised our dream of existing as things, our dream to be upstarts in the world of products. There is in fact no fundamental ontological difference between the star actress, disseminated by thousands of photographic-replicas, and nail varnish that circulates in countless examples. It is entirely logical that stars and products form an alliance and support each other in adverts (the star by recommending the merchandise, the merchandise by indulging images of the star): they are 'birds of a feather that flock together'. Not only are they equally well known, they have also overcome their mortality in the same way: both can continue to prove themselves after their death in the form of their reproductions. Like nail-varnish that can continue its path and keep on shining once the original prototype has ceased to exist, the smile of the star can continue to shine after the original has gone the way 'of all flesh'. In a certain sense, the star is even 'immortal whilst still alive' (the 'immortal Garbo' for example) and is not bound by the way of all flesh: she is younger than she is, because the majority of her pictures feature the eternalised version of her actual self (i.e., the only version that counts commercially), the one that portrays the divine and wrinkle-free appearance from her youth. The way of her actual flesh is a process that no longer counts and that can at best only be shamefully concealed. 'Hollywood-privacy', an entirely new type of 'privacy' is in the process of taking shape. It seeks to hide everything that could become detrimental to the validity of images.

Nevertheless, in comparison with the truly multiple existence that our mass products enjoy, our own multiplication through images is destined to remain an 'as-if'. Despite our intense iconomaniac activity, the satisfaction we gain from this is derived from an inadequate substitute. The difference between actual examples and mere copies, the fact that we have to settle for multiplying ourselves in the form of photographs, whereas products are allowed to disseminate themselves across the world in truly identical pieces, cannot be eliminated. Humans can still not completely extinguish the shame they experience in front of their 'superior things' through the images they have of themselves.

So does this mean – one will ask – that in their most deeply held dreams and desires humans see themselves as actual mass products? That they positively desire to live as mass products?

It is not possible to conclusively answer this question in the affirmative, for the same reason that it is impossible to assert that humans positively long to be for ever – in the sense of '*sempiternitas*', and not '*aeternitas*'. It would be very easy to see in our fear of death and our precautions against dying evidence in support of this assertion. But this reasoning would be misguided. Certainly, it is true that we *do not want to die*. It is not true, however, that we positively want to continue to live on and on, that we want to become countless millions of years old. We are not even capable of 'imagining' such a forever on and on. This means that the certainty is exclusively negative and precautionary, not a positive desire. We cannot pursue here why it is that no positive mental image can match this most fundamental desire to negate death, why our imagination remains *blind* and why we are in a sense 'too mortal' to even positively '*think*' that we will never die. But that's the way it is.

Our desire for multiple-existence also remains 'blind' in the following sense. It is true that each of us wants to divest of our 'unique existence'. It is not true, however, or at least it is a 'philosophical exaggeration' to suggest that everyone shares the desire of the Californian patient, to divide himself up into 'spare pieces' so he can populate the world in the form of many or even countless examples. This is why our 'iconomaniac compromise' that consists of taking part in the mass-existence of products through our images while remaining ourselves might be the best solution after all.

10

An illustration of humiliation sourced from history: MacArthur as a precedent

Following the analyses offered so far, in which we consistently employed anonymous phenomena of our age to illustrate the symptoms of 'Promethean shame' and the lengths to which we go to combat this feeling, we want to portray at least one case in which this *humiliation* is visible *as an identifiable historical event.* It is evident that the outperforming of humans by machines that puts them to shame must be noticeable as *instances* of *degradation* when humans are in the public eye and are on the highest rung of the social ladder. MacArthur was a human being of this kind.

For many years it has been an open secret known around the globe that at the beginning of the Korean War General MacArthur proposed measures that arguably could have triggered a Third World War. It is also common knowledge that the decision as to whether such an outcome should be risked or not was taken out of his hands. Those who removed this responsibility from him, however, did not do so in order to take it upon themselves, or to pass it onto other people with political, economic, or moral expertise. Rather, they removed the decision from MacArthur to hand it over to a *machine* (for the 'final word' in this matter had to be objective – and today only statements made by objects are deemed 'objective'). In short, they 'transferred the responsibility and answerability'* to an '*electric brain*' which they considered to be the highest and final authority in this matter. The decision was thus not merely removed from the specific person MacArthur, but from him

* The German is '*über-antwortet*', again it possible to hear a resonance with Heidegger's vocabulary. *Being and Time* presents human existence as unfolding in the space opened by the experience of 'being delivered over and made answerable' (*Überantwortung*) to the inescapable fact that one 'is and has to be' (see esp. the discussions of 'thrownness' and 'facticity'; §29ff.) – Trans.

as a human being. The machine brain was pitted against MacArthur's not because there were particular reasons to question his intelligence, but because of the simple fact that MacArthur merely had a human brain.

It is, however, misleading to say that this responsibility '*was* being taken away' from him as a human being. After all, the power wresting the decision from MacArthur was not some superhuman authority, no 'Moira', 'Tyche',* 'God' or 'history', but *humanity itself.* In a sense, humanity used its right hand to rob its left, offering up the booty – its own conscience and freedom to decide – on the altar of machines. With this, humanity proved that it had submitted itself to this man-made calculating robot and was willing to accept this machine as a substitute for its own conscience and acknowledge it as an oracle-machine and even as a machinic eye of providence.

This is to say, therefore, that *humanity harmed itself* by subordinating the General to this machine. One should not misunderstand me here. This is anything but an expression of solidarity with MacArthur. That he was placed centre stage as the *dramatis personae* in this exemplary moment was a mere whim of history, any other person in the public eye would have served history just as well. What we are alluding to, then, is merely that anyone who transfers responsibility from a human being to a machine also transfers their own responsibility. And with MacArthur, humanity practised this form of self-degradation for the first time in a representative manner. For the first time it publicly said to itself: '*Because we aren't as good at calculating* (rechnen) *as our machine we are unreliable and incapable of taking responsibility* (unzurechnungsfähig)*, so let's not "make decisions"* (rechnen) *anymore.*'† *For the first time humanity didn't feel any shame at publicly humiliating itself.*

This oracle-machine was 'fed' with all the data about the American and the enemy's economy – *N.B.* 'to feed' is the technical term for supplying machines with the data they require to reach decisions.[23] However, it is an exaggeration to say: 'with all' the data. For a quintessential aspect of machines is that they have an '*idée fixe*', that is to say, they impose determinants that are limited artificially by what they can do: the fact is that the perspective from which data is computed is regulated in advance and remains fixed in this manner once and for all. And so they 'fed' the machine exclusively with the type of data that did not offer any resistance to quantification. Datasets, therefore, which merely touched upon benefits and damaging effects, the profitability and unprofitability of waging the war that was being considered. Of course,

---

* *Moira*: The Goddess of Fate; *Tyche*: The Goddess of Chance and Fortune – Trans.

† The text is here playing on divergent strands of meaning the word '*rechnen*' can carry: (1) 'to calculate, count and account'; (2) to 'anticipate', to 'weigh up', to 'process' and 'work something out'; but also (3) to 'rely on' or to 'count on'. Someone who is '*unzurechnungsfähig*' is of 'unsound mind' and therefore declared 'not responsible' and 'not liable' for his or her actions. Today, there is an added irony here because one of the German words for a PC is *der Rechner* – Trans.

for reasons of methodological unambiguity and clarity, this procedure had automatic consequences; for example, the annihilation of human lives or the devastation of countries could only be considered and evaluated as figures of profit or loss. Questions like 'is this a just or unjust war' were not even put to the 'electric brain'. In fact, they would have even been embarrassed to dish up such fare, for it was foreseeable that the machine, in its incorruptible object-objectivity, would have refused such sentimental and subjective fodder. Or, if they had nonetheless violently stuffed such questions down the throat of the oracle-machine, it would have not digested them and would have responded with an electric constipation. In short, out of consideration for the machine's exalted allergy towards the subjective, they renounced moral questions in advance. Following the model of 'logical empiricism', they only classed questions as 'sensible' to which the machine could provide an unambiguous answer. They rejected all other types of question as nonsensical (and this even though they tried to convince themselves that the machine had been consulted in the first place only because of acute moral scruples).

Two things are thus sidelined, are no longer 'at work' and no longer 'count' (*rechnen*) when recourse to such a machine has been found:

1. The competence of humans to resolve their problems themselves, because in comparison to the machine their ability to calculate and work things out equals zero.
2. The problems themselves if they cannot be evaluated and computed by a machine.

As is well known, it takes a ludicrously short time for data to be processed by a mechanical digestive system. They had hardly finished feeding the apparatus when it excreted its oracle. And since they were willing to accept this excretion as a verdict, they knew at this point whether they should dare the undertaking and announce a holy war, or write it off as loss-making business and scrap the whole idea as immoral.

After a few seconds of profound electric reflection or digestion the 'brain' excreted a decision that by chance was more humane than the decisive-strike the human MacArthur had proposed. The 'brain' resoundingly proclaimed that MacArthur's approach would be 'loss-making business'; yes, it even portrayed the potential war as disastrous for the American economy. All this was certainly very fortunate; possibly even the best fortune humankind has ever had, for the oracle sounding from the mouth of the machine was already being carried into the atomic age. However, the process, as such, by which this decision was reached, was at the same time the most epoch-making defeat that humanity could have inflicted on itself. For never before had humanity degraded itself to such a degree that it entrusted judgement about the course of history, perhaps even about whether it may be or not be, to a thing. It is irrelevant that the machine's verdict on this occasion was a veto and a reprieve, for it still constituted *a death sentence, precisely because the source of possible reprieve had been transferred to a thing.* The status of humanity was not decided by the machine's positive or negative answer but rather by the fact that this question had actually been submitted to the thing and the answer abided by. If one realises that thousands of humans, and we are possibly amongst them, owe gratitude for their chance of not-being-dead-yet – this is what counts as 'life' today – to the 'no' excreted by a machine, then shame would make one want to sink into the ground.[24]

But let's return to MacArthur.

His proposition has been consigned to the oracle-machine for evaluation and arbitration. Even before the machine has voiced its support or its veto, the '*right of decision is taken away from*' the general.* MacArthur has to accept what the machine utters as the final word. Even if it had given the go ahead, he would, of course, still have lost 'the right to decide': for what would be left for him to do after the machine's announcement, could only be done because of its electrical permission. The humiliation has hence begun.

But as we know, the proclaimed verdict was a *veto.*

With this the humiliation is complete. For a man with MacArthur's ambition and imperiousness the situation is untenable. The consequence cannot be avoided: although he is not demoted in the military and technological sense, his demotion is nevertheless a fact. His military career has come to an end. He becomes a civilian.

This was the first consequence. It is logical and bears no surprises.

But the story of MacArthur's humiliation is not over yet. There was a second consequence, which, if I am not mistaken, constituted a *countermeasure.* And this countermeasure is anything but ordinary, it is highly astounding. For, this story reached a climax that would more likely be expected from a novelist rather than reality: MacArthur became the president of a corporation making office machines.†

It is, admittedly, not impossible that this move was a coincidence, that amongst the countless job-offers MacArthur received on his return to civilian life this was the most attractive. But this seems unlikely. We believe, therefore, that we are at liberty to portray this climax, at least on a trial basis, as the product of an *action* that *means* something. We will do so with every

* The German here is '*entmündigen*', meaning to be stripped of being '*mündig*' ('to be of age, to be responsible, to have the right to decide'). The word '*mündig*' makes reference to the word for mouth ('*Mund*'), thus implying that when coming of age, one is fully 'responsible' for what comes out of one's mouth – Trans.

† In 1952 Douglas MacArthur joined Remington Rand, the company then producing and developing the UNIVAC computer – Trans.

reservation and will leave the question open, as to whether we are staying within the boundaries of fact or whether we have already leaped into the realm of 'Science Fiction', or rather 'Philosophy Fiction'. Let's spell this out.

If we were to portray MacArthur's choice of career in a novel, then we could only explain his motivation as follows: unable to come to terms with the humiliation the machine had inflicted on his authority, MacArthur was looking for ways and means to rid himself of the blemish. Under these circumstances, the executive position in a company that manufactures business machines was the only position that would allow him to gain compensation. It was the only position in which he could show the machines that his voice and verdict still had weight, and prove to them that they too were still under his dominion. It was only here that he could master and humiliate the arrogant machines, at least, in business and administrative matters. It was only here that he could quench his thirst for revenge. And that's why he chose that particular profession.

This motivation could be translated into the dialectical language of Hegel without any difficulties. In a version of *The Phenomenology of the Spirit* written for today one would then read:

'After the former "slave" (the machine) had become the new "master" by demoting the former one (MacArthur) to the rank of a "slavè," this "slave" attempts to become once again the "master of the master" (the machine)'. This is the story of the humiliation, at least the hypothetical one.[25]

11

Shame as a disruption of identity – the expression 'ontic endowment' – The 'I' is ashamed of being 'itself'; the 'it' is ashamed of being an 'I'.

In conclusion I must engage with the strongest objection that was raised against my thesis of 'Promethean shame'. Contesting this most fundamental argument requires such an unusually extensive 'digression' that it was not advisable to bring it in the course of the investigation so far. The objection in question tries to demonstrate that the whole theory is unfounded and claims that the word 'shame', as used in the expression 'Promethean shame', is *only a metaphor*. 'Real shame', the objection explains, is always felt in front of an authority acting with a superior power and whose surveilling gaze is feared. Now in the case of 'Promethean shame' (if there were such a thing) the authority would be the world of machines, and would therefore consist of sightless things. Because these cannot see the imagined human flaw, and consequently cannot evaluate and condemn it in any way, the whole thesis is delusional. The word shame it deploys amounts to nothing more than a metaphor and, while one may well allow the passive verb form that suggests that humans *are being 'humiliated and embarrassed'* by the sophistication and multiplicity of manufactured things, this is not the case with the self-reflexive verb form suggesting that humans are actively 'ashamed of themselves'.

If we are to meet this fundamental point of contention, we first have to understand the nature of 'shame as such'.

What, then, *is* shame?

I.

a. A reflexive act (to be ashamed *of oneself*), that is, it involves a *reference to self.*
b. Shame, however, is a reference to self that *fails.*
c. This reflexive act does not fail only occasionally (as it does in other self-reflexive acts, such as remembering oneself); it *fails categorically and fundamentally.*
d. This failure is categorical, *because those who are ashamed encounter themselves at once as identical and as not self-identical* ('It's me, yet it isn't me after all').
e. As a consequence of this, the *act never comes to an end*: because the one who is ashamed cannot 'get over' the contradictory encounter, shame is never over and done with either. (In this and the traits f and g, shame resembles astonishment).
f. So much so that it actually forfeits its quality of being a reflexive act and degenerates into a state.
g. It does not, however, become a state of equilibrium, a stationary 'mood', but an oscillating state of *irritation and of disorientation*; it turns into a condition that seems to perpetually begin afresh, even if one already deems oneself to be in the thick of it. In brief: *shame is an interference in processes of self-identification*; it is a condition of 'being distraught and confused' (*Verstörtheit*).

II. In contrast to most of the harmless 'acts' that usually are the objects of psychological and phenomenological investigation, shame essentially entails a '*double intentionality*': it is not solely directed towards its (normal) 'intentional object' (the perceived 'flaw'), but is also always simultaneously turned towards an '*authority*' *before* which the one who is ashamed feels shame. It entails a '*coram*'.*[26]

III. This authority is an *unsolicited authority*, often even one which is unwanted; it is hence not actually 'intended', but is rather evaded. The intentional turn towards this authority is in fact a turn away from it.

* The Latin *coram* means 'to be in the presence of', 'to be before', for example, the eyes of the public, the law or an authority – Trans.

The intentional reference to it is a rejection, that is, it is '*negatively intentional*'.[27]

The traits listed in section I will be the main subject of our investigation. For this reason I will quickly make some comments about points II and III first.

Neither '*coram*' nor 'negative intentionality' have been given any specific importance in phenomenology.

ad II: '*Coram*' is not a trait specific to shame, even though it is particularly evident in this case; it is a trait that belongs to all social acts. (Example: 'to boast in front of someone'). '*Coram*' is probably an element that can be discovered in every act, however hermit like it may be. Even if one is oblivious of this, even if the act takes place in private, it is doubtful that there is an act that does not make reference to a particular social-world (*Mitwelt*) on which it counts and before which it takes place or wants to pass unobserved. The conception of consciousness in the classical phenomenology of Husserl always had to closely border on solipsism, because this phenomenology, without being clear about its own principle of selection, almost exclusively restricted itself to the analysis of acts that did not openly imply this '*coram*'.

ad III: '*Negative intentionality*' remained unknown, because the 'ego' was assumed to be *eo ipso* 'free', or at least to be '*active*'. The fact that the word 'freedom' was seldom used (e.g. in Husserl) does not say much to contradict this. No different from other philosophies of subjectivity, 'transcendental phenomenology' is marked by a classical lapse. In its pursuit of knowledge of the 'ego', it only captured the tip of the ego, the 'cognitive, perceiving I', and mistook this for the whole ego or as representative of it.[28] It is undeniable that the 'cognitive, perceiving I' pursues an object, that it is 'positively intentional'. But that the ego is only ever 'pursuing something' is not true. Besides, the ego also withdraws into itself (in the manner of a snail) and intentionality only gains its meaning in coordination with this countermovement; the ego does not only see, it is also seen; it does not only intend, it is also the object of intention, for example, when it is pursued or chased. It is highly questionable whether the object someone is fleeing from can be called an 'intentional object' in anything other than a grammatical sense. After all, not everything that is grammatically an accusative is automatically also an 'intentional object'. Now shame also belongs to this class of two-way intentional acts, because it turns its back on the authority in front of which it feels ashamed. Shame neither intends to see or not to see the higher authority.

What it wants is not to be seen. This relationship differs so fundamentally from what is usually called 'intentionality' that one ought to actually invent a new term to describe it.

If one sums up the traits mentioned so far, then shame is: *A reflexive act that leads to a state of confusion. This reflexive act fails because in the process humans experience themselves in front of an authority from which they turn away. They experience themselves as something 'they are not' in a manner that inescapably impresses on them that 'they are this after all'.*

After this academic-phenomenological description, which still leaves the fact *that* shame exists unexplained, it will be good to jump to a concrete example.

Example: The hunchback is ashamed of his hump. (More correctly: He is ashamed to be the one with the hump.)

In a certain sense, the hump appears to him as something contingent, that is, not as what he 'is', but only as something that he 'has'. *What one 'has'*, however – 'has' in the way that one 'has' one's body – *one 'is'*, and this inescapably so. Hence, the hunchback is inescapably also 'the one with the hump', with whom he now has to identify; even though he cannot do so; even though he 'can't help' his physiology. Because he cannot get over this contradiction, he cannot get over his shame: it does not abate.

As for the 'authority' in front of which he is ashamed: it consists of a *tribunal that decides how humans actually 'ought to be'*. He too, belongs to this tribunal. He too approves the human-canon this tribunal represents; no less so, in fact, than the members of the tribunal with a more fortunate build. Because he accepts the judgement of the tribunal, but then again, as the misshapen one, he cannot let this judgement sink in (and he cannot want to let it sink in), he turns his back on the tribunal – and also turns away from himself. Here too, a 'crisis of identity' is evident.

Moralists will find this basic example repulsive. 'If he can't help the fact that he has a hump', they will exclaim with moralistic blinkers, 'why does he need to be ashamed of it?' The response to this interjection is not that the hunchback has not also thought of this ingenious line of argument. (Quite the opposite; as long as he can remember he has been telling himself 'I can't help it – my body is not my fault' and he never tires of trying to extinguish his feeling of shame in this way.) No, the answer to the moralist's interjection is that these great efforts remain futile because he does not care about the question of whether he 'needs' to feel ashamed or not. In short, he is ashamed precisely because of his inability to do anything about it. *The hunchback is ashamed of his physiology not despite the fact that it is not his fault but because of this fact.*

As it happens, no other saying points to what shame is as unambiguously as: *'I cannot help it, it is not my fault'. Everything that I 'cannot help', everything that I 'cannot do anything about' is beyond the reach of my freedom and self-determination; as such, it belongs to the realm of fate and is 'fatal', a measure of 'impotence', in the broadest sense.* Shame has its source in the contradiction that exists between the pretention of freedom and the 'fatal'; between ability and inability. Shame is ashamed of inability. What does this mean?

The aspiration to freedom by its very nature knows no scale and is without measure: to be partially free is not enough. The self (*das Ich*)* does not only want to be itself now and again; the individual does not only want to be individual as an attribute. The self, rather, desires to be absolutely free, individual through and through, nothing but itself.

This overstretched aspiration is 'pathological'.[29] It cannot be sustained in the long run. There is always a point at which the self comes up against the limits of its own freedom, individuality and self-confidence. In such moments, the self finds 'itself as something' that it is too, even though what it finds is not identical to what it is as a self or as an individual. This is the moment when the self discovers that it also exists as an 'it'.

With the word 'it', I do not only mean what Freud called the 'id' (*das Es*), but something much more general: everything that is not of the self. The 'it' designates everything pre-individual, whatever this may be, everything that the self is part of without being able to do anything about it. The 'it' denotes what is '*given*', what the self *also* has to be, insofar as it is. That's why we address it here as the '*ontic gift*'.[30]

Shame is generated in the instant 'something given' is discovered. I say explicitly: 'in the instant'. Shame does not limp behind to set in after the event. Shame is not to be understood as a reaction to the realisation of one's inabilities. Nor is it a statement (*Stellungsnahme*) about this state of affairs (as sadness is, for example). Shame is this inability, shame is this failure as such. This makes shame incomparably more fatal than sadness. When, for example, an ascetic finds himself confronted with the fact that he 'has' a body (and 'to have' here also means 'having to have', 'not to be able to do anything about this circumstance', hence, that one is passively 'being had'), when he is confronted and perplexed by this fact which is as unacceptable as it is undeniable – then this bewilderment and perplexity is his shame. When the 'self' demanding freedom has to realise that it was not posited by itself (in Fichte's sense), but that it rather 'became', that it would not exist as a 'free self' at all, if it were not at the same time arrested by the 'endowment' of an

* In the aid of stylistic clarity, I have used 'the self' to translate *das Ich*. 'The I' or 'the ego' would have been more literal alternatives – Trans.

unfree, conditional and creaturely humanity – then this helplessness and this failure is the shame of the self. 'To be ashamed' hence means: '*Not being able to do anything about the fact that it is not one's fault*'. Put another way, the very thing on which the moralist stakes his claim for an absolution from shame is precisely its generative moment, its basic motif.

What holds true for the hump of the hunchback, the body and the origin, also applies *mutatis mutandis* to the 'bad deed': because one is primarily ashamed of *something that is* (eines Seins), of *having* something, and not because of what one has done, the latter cause of shame is not nearly as fundamental as the hunched back in our example. Ultimately the perpetrator is not ashamed of his misdeed, but of being 'the one who' *is* identical to the one who committed the bad deed. In brief, he is ashamed of *something that is*.[31]

Shame erupts because one is simultaneously 'oneself' *and* something else. At the same time, it is in a sense also an attempt to shed this 'other quality', the 'endowment'. We have already seen that this attempt is futile.[32] In fact, shame is so perplexing and so without method that it leaves the person feeling shame no other escape route than the defeatist one of going under together with the shameful 'endowment' (to which he or she is chained). The figures of speech: 'to want the ground to open and swallow one up' or 'to sink into the ground' are not metaphors. They are accurate descriptions that cannot be replaced with more 'exact' ones. And yet we know that this obsession with disappearing is of no avail, because the ground petrifies under the feet of shame. And now shame begins to accumulate frenetically as 'shame about being ashamed'.

Once one has grasped the basic condition of the one who is ashamed, and has understood that this disorientation and despair is brought on by the recognition that freedom and individuation are limited, then it is evident why *sexual shame* has time and again ended up being regarded as *the quintessential* form of shame. Sex and that which pertains to the genus is the pre-individual sphere *par excellence*. Its realm lies beyond choice and freedom and as such it provides the pre-eminent 'it', which does not belong to the individual *qua* individual.

For the individual, the claim that its sex does not belong to him or her remains a hollow claim: for a bond exists between them, and in two senses. On the one hand, the individual 'belongs' to a sex (and is subject to 'longings'). On the other hand, sex can only actualise itself in an individual. The bond is thus so close that it does not make any difference if one asserts 'the individual is an attribute of its sex' or 'sex is an attribute of the individual'. The truth lies in this ambiguity, in the simultaneous validity and nullity of these two statements.

This ambiguity, or, this 'dialectic' fact if you will, is just the reason why one's sex is a pudendum, why it is something of which one is ashamed. Inasmuch as individuals are sexed-beings, beings who 'belong' to a sex and who 'long', they are not master of themselves and are hence not 'themselves' and not free. In a most ambiguous manner they are *at once themselves and not themselves*, and thus the formula according to which we have begun to conceive shame is confirmed.

Viewed from this perspective, shame is thus not any random psychological state (insofar as any such states exist at all). It is rather a *methaphysicum*, the embodiment of the dialectic of 'the particular thing' (*res*) and 'and the universal', a dialectic that has long been the object of contestation in universalism. The '*res*' (here: the individual) is ashamed of embodying the 'universal' (or at least the general – the individual is ashamed of being its attribute, of being it). It is not without reason that the one 'thing' that all have in common, the 'shared feature' incorporated in every single body, is called the '*private parts*' ('*die Scham*').

If one restricts oneself to the example of sexual shame, then our definition of shame is confirmed: shame is a disorientation of human beings who are forced to find an 'it' instead of an 'I' and laboriously and futilely try to identify themselves with themselves completely. But does this hold for other forms of shame? Will they confirm our definition just as well?

Let's put this to the test.

The shame of the shy child is surely no less elemental than sexual shame. When a child hides behind its mother's skirts, does it do so to conceal its existence as an 'it'? Does the child hide because it feels cheated out of its entitlement to only be an 'I' (or exclusively a 'self')?

No.

Does it even make a claim to such an entitlement?

Likewise no. It seems pointless to attempt to establish even the most artificial link between this example and our previous definition. Our classification of shame as a 'disturbance in processes of identification' seems to contain a mistake.

But where is this mistake? Let's engage in a short reflection.

'Identification' requires two parties: the *one who* wants to identify – we will call him or her the '*identificator*' – and the '*identificandum*', the thing *with which* he or she attempts to identify.

So far we have obviously limited ourselves to cases in which the 'self' was assigned the role of 'identificator', whereas the general and impersonal realm of the 'it' was the 'identificandum': *the 'self' was ashamed of the 'it'*.

But is this the only possible distribution of roles? Is an inversion of this dynamic not equally possible? Is it not conceivable that the 'it' could end up in a situation in which it was forced to identify with the 'self' being imposed on it? And would this imposition of identification not lead to disorientation? Precisely to the disorientation we call 'shame'? In brief, is it possible for the *'it' to be ashamed of the 'self'?*

This question sounds strange. The child we are summoning now will immediately convince us how little this is the case.

The existence of the child is still embedded into the restrictive ground of the family, limited to a mere togetherness, a 'being-with-others' ('*Mitsein*'). The child has not yet developed into an 'I' and has not yet individuated into a 'self'. The thought of '*taking itself* from the ground in which and with which it lives has not yet entered the child's mind.

I say: '*take itself from*' (*herausnehmen*). Initially this seems to mean nothing other than 'to take something from something else and to make it an exception (*Ausnahme*) by doing so'. But because the turn of phrase 'to take oneself from' (*sich herasunehmen*) already shimmers with the connotation 'unrestrained impudence', we can expect a certain relationship between 'being-oneself' and shame, if one can trust the wisdom of language.*

Let's assume that this child is addressed by a stranger who asks the child's name, for example. This customary opening question can stand in for any other question that might be asked, as it requires that the one addressed '*take*' the step '*out*' of the group by saying their name and identifying themself as themself.

The status of the child has been transformed by a single blow: instead of still only existing 'with-others' as a mummy's child, instead of remaining indistinctly in the background, it now has to stand its ground as an 'exception', as a single being who is no longer part of the ground it was taken from. The shock of this social birth is violent, so strong, that it can perhaps even be understood to be a variant of what Freud calls the 'trauma of birth'.[33]

The child is of course not entirely ignorant of this new state. Although it had not yet 'taken the liberty to separate itself' from the ground in which it was embedded, it had already been exposed to the danger of 'being-itself' from the moment it was born – the birth itself was the first 'abandonment': every child is an 'abandoned' child. In its hours alone, particularly, when 'left to itself' in a state of deprivation, it had to experience for the first time existing as a self: it had to come to terms with being without being-with-others. Since anxiety is the engine of insight, the child had already begun to recognise, or at least to intuit that the blissful indistinctness of existing exclusively with others would one day end for good, and that a colder, more exposed mode of being would become its destiny.

But *who* had this intuition? *Who* was the subject at the heart of it, who recoiled anxiously?

The self?

---

* The text is here playing on the shift in meaning the verb *herausnehmen* ('to take out of') undergoes when its reflexive form is deployed: *sich etwas herausnehmen* means 'to take the liberty to (do something)' – Trans.

Is it conceivable that it is the self that is recoiling, when the thing making it recoil is the looming anxiety and dread of being a self?

Inconceivable.

As strange as this may sound, the only conceivable answer to the question who the recoiling, intuiting subject could be, is the being not yet fully formed into a 'self', namely, what we have called the 'it'.

And this catastrophe now unfolds. The very instant the stranger addresses the child the threat becomes real. Even though the child is still an 'it' [without a distinct 'self' – Trans], it sees itself challenged and called forth as an 'I'. Because the child cannot yet meet this challenge, too much is demanded of it. That the child already knows and recognises '*I* am being addressed' or even 'it *is* me' does not even have to be disputed. It is equally certain, however, that the child also knows 'it *is* not yet me'; '*I* am not yet a self'. In brief: *the situation is a concurrence of being-a-self and not-being-a-self.* This paradox is no less indissoluble, no less indigestible and no less unsettling than the situations we have recognised as shame inducing above, situations to which there is 'never any closure'.

The 'it' now really has no other options than the disturbed ones described: overwhelmed by the anxiety of having to emerge as a 'self' in public, the child only knows the desire to sink into the ground and to become invisible. In the manner of a kangaroo that jumps back into the pouch from which it came, the child attempts to regain the alibi of merely existing-with-others and hides in its mother's skirts.

Schelling said, and this expression was already implied above, that 'the arrogance of man rises up (*sträubt sich*) from the ground against this origin'. We may now also add an accompanying sentence: 'The origin is reluctant to take the leap into freedom and into the exposedness of existing as a self'. In this refusal and reluctance there is also 'shame' – and probably not only 'also': there is every indication that the shame of being 'oneself' is the more common and originary form of shame than the feeling of shame about not being oneself.[34]

That's right, the more general and originary form. Why, after all, and because of what, does one usually feel shame?

A speaker who has to confront a crowd (instead of remaining comfortably concealed within it): is he or she not ashamed of appearing *as themselves*? Is he or she not ashamed of being abandoned to 'the public' under a barrage of looks? And what about someone wearing unusual clothes or a foreigner in foreign lands? Do they not run the gauntlet because 'they take the liberty to stand out' and because they are 'exceptions'? They attract attention because they *are being themselves*, and as a consequence they contrast with others and stand out. And does not everything indicate that the shame of breaking norms is the prototype of *moral shame*? The disobedient who 'take the liberty' of standing outside of the normal and the norm, are they not ashamed when *exposed as being themselves*? There is here a stark contrast to our earlier examples. These now begin to lose some of their relevance. Do those who are ashamed not generally conceal themselves because they perceive their *selfhood* as a flaw? When they 'voluntarily stand in the corner', or when they pretend not to hear, do they not all attempt to extinguish their shameful conspicuousness by denying that they exist as a self in order to reassume their old status prior to their lapse?[35]

This digression has delayed us for a long time. But since this seems to have given us some insights into what shame is, it has left us better prepared to face the objection that originally gave rise to it.

The objection was that the expression 'Promethean shame' was purely metaphorical, and that, therefore, it would not allow us to see true feelings of shame in the experiences the expression describes. The reservation was that the 'authority' provoking these experiences could not be a true, shame-inducing authority, since it consists merely of blind machines. The thought that humans could be eccentric enough to accept eyeless objects as true authorities, or even that they would be afraid of these objects as 'awe inspiring powers' (*tremenda*) that can make them shudder and tremble, is simply absurd. This was the objection.

Clearly we are now left with two tasks: (1) We are obliged to examine the accuracy of this objection – an examination, which in itself has two components: (a) the question of metaphor; and (b) the question of 'eyelessness'. (2) We need to examine – and this is the main task – whether 'Promethean shame' represents 'a disturbance in processes of identification'. Since we have defined shame as such an interference, only this will allow us to decide whether 'Promethean shame' can be classified as a 'true form of shame'. With this we have sketched the path of our concluding analysis.

12

Dismissal of Objection (a) – No expression is purely 'metaphorical'.

We will begin by responding to the objection that we are letting ourselves be led astray by a metaphor.

What is the point of this objection? What does the claim mean that a description of a mental and emotional reality is 'exclusively metaphorical'? Is there even such a thing?

Are human beings that we call 'bright' only 'bright' in a figurative sense? Is a 'heavy heart' (*Schwermut*) only an allusion, because the true literal sense of 'heavy' (*schwer*) is deemed to be an exclusive attribute of physical objects?* Why exclusively of these? Where is it written that the word 'heavy' was originally deployed to describe physical qualities?

When psychologists discredit the diagnostic value of such expressions, when they admit in a voluntarily gesture of self-limitation that they know that for a fact the employment of such metaphorical expressions leads to very little that is rigorous or worthy of being taken seriously, then they do so primarily because they are intimidated by the rigor and the authority they accredit to their bigger sisters: the natural sciences, and not due to real methodological concerns. And here, they are in the wrong. For, stripped of their linguistic dimension the mind and feelings are entirely without essence: the way in which psychic reality represents itself linguistically is part of this reality itself. When something 'weighs on' the soul, when it describes itself as 'weighed down', when it recognises itself in this word and is able to convey this to fellow human beings, then this is evidence that the 'heart' is truly felt to be 'heavy'. The idea of seeing expressions such as 'heavy' merely as a label that was peeled away from another (physical) source and stuck onto 'heavy heartedness' is rooted in the unspoken presupposition that the soul invents its forms of expression in the same arbitrary manner in which industry comes up with names for branded goods. The allusion to 'weight' is in fact a real component of the emotional disposition felt. And even if one were to admit that the term 'weight' is a label that has been removed and stuck on elsewhere, the very fact that the soul has reached for precisely this label and not for another to characterise its own condition would signal that it actually recognises itself in it.

Put otherwise, the suspicion directed at metaphorical expressions is based on an erroneous notion which is nevertheless taken to be self-evident. This is that different realms of sensation and experience are autonomous and hermetically sealed from one another and that competent people, namely scientists, are not permitted to engage in small scale border crossings between these realms. To prove that such a psychology is absurd and that the expressions imported from allegedly foreign areas are readily understood is not really necessary. On the contrary, the very fact that expressions imported from allegedly foreign provinces are so readily understood demonstrates just how widely open the borders are. In everyday life no scientist can get by even for an instant without having recourse to the very metaphors that he discredits. And he cannot contest that metaphors are an excellent resource for communicating things that resist 'non-metaphorical' formulation to others in a clear manner. This state of affairs should make us suspicious, but not of metaphors – every beginner who has heard anything of the rigour of science has

* The word *Schwermut*, which describes a melancholic, gloomy or wistful emotional state, contains a reference to weight (heaviness). I have tried to preserve this by translating *Schwermut* with 'heavy heart' – Trans.

this suspicion anyway. No, it should make us *suspicious of the legitimacy of our suspicion of metaphors.* To complain that we cannot access the reality of the psyche because of our 'merely metaphorical' language is misguided; *on the contrary, precisely in recognising metaphor we grasp one of the most essential traits of the soul.*

When people who are ill moan to us about a 'dull' (*stumpf*) pain, then we understand exactly what they feel and what they want to say. An imaginary doctor who dismissed this expression as 'unscientific' or as 'merely metaphorical' would be a character for a farce about our scientific age. Especially if he were to accuse the patient of using an expression that belongs to another realm of meaning and that on account of this the patient was guilty of a μετάβασις εἰς ἄλλο γένος.* That the sick person is 'trans-ferring' (μεταφέρει) the meaning of the word 'blunt' (*stumpf*)† from another physical realm to his pain is not at all true. The expression 'blunt', or rather the quality this expression points to, does not only inhabit one specific realm. Or put positively, the expression denotes something 'pre-specific'. It portrays a relationship between the world and the subject that does not yet have anything to do with the division of the subject into different sensual realms. The specific actualisation of this quality in a particular sensual realm is always secondary. Only with this secondary motion does 'the blunt' become something that is blunt to the touch, or blunt acoustically, osmotically or emotionally. If this were not the case, we would not be able to comprehend the words used at all (and the language of poetry would remain obscured from us). We can only apprehended what a 'dull pain' (*stumpfer Schmerz*) is, and even what a 'blunt knife' (*stumpfes Messer*) is, because we comprehend what 'bluntness' is in a pre-specific manner.[36] And what is true for 'bluntness' also holds for all other experiences that are actually or allegedly characterised with 'metaphoric' expressions, such as 'exaltation', 'relief', 'feeling overwhelmed', and 'depression'; and of course this is also true for shame.

13

Dismissal of Objection (b) – The world is not without eyes.

So much for the subject of metaphors.

My opponent had offered the following reason for his accusation that my use of the word 'shame' was metaphorical: One can only speak of real shame

* μετάβασις εἰς ἄλλο γένος: a step across into a different area of meaning – Trans.

† The word *stumpf* does not translate smoothly because English makes a distinction between 'dull' and 'blunt' that German does not. The paragraph is bringing together the sensation of handling 'blunt objects' and feeling 'dull pain' – Trans.

when those who are ashamed know themselves to be under the watching eyes and surveillance of an 'authority'. The world of machines *does not have eyes*; no human being is eccentric enough to assert that machines can see them. It is, therefore, out of the question that machines could function as 'authorities' of this kind.

This seems plausible, but only because we are biased by theories and no longer know what is plausible to someone who is not. To those who are unbiased, nothing is more persuasive than the notion that they are being looked at by things. Such humans are anything but theoreticians of epistemology and cognition (*Erkenntnistheoretiker*) who conceive of themselves merely as observers (of the world). No, they know with the same level of certainty that they are also seen and observed (by the world). And this does not just mean that they know themselves seen by fellow humans and animals, but by the whole visible world. Normally, at least originally, humans deem 'visibility' a reciprocal relationship: *Everything that they can see also sees them.*

It is unnecessary to emphasise that we are not asserting that humans are actually seen and observed by a chair, table or picture; what we are suggesting is only that with an unprejudiced 'worldview' humans consider themselves 'viewed by the world'. They do not make this into a theory, precisely because this is a 'pre-theoretical worldview' that dissipates the very instant a theoretical attitude is assumed. That is also why the term 'worldview' might be inappropriate here. At any rate, the use of the word is irrelevant in view of the fact that we *behave in a way that suggests that 'we are seen'*. This is more conclusive than any theory which could be laboriously deduced in retrospect. This claim could be supported with countless pieces of empirical evidence. For instance, by the material compiled – albeit with a completely different aim – by psychoanalysis. This provides examples of how the inhibitions of shame are still at work in completely undisturbed isolation in a natural environment. Robinson did not walk around naked either.

We also readily allow poets to tell us that the mountain peak is 'facing us' with menace, or that the moon is 'looking down on us kindly'. And we do not reject such information as eccentric, but understand what the poets mean to say with it. This shows that we still take 'seeing things' for granted. We, too, see something in their '*look and appearance*' that also makes things '*look at us of their own accord*'. In the idiom of poets we thus rediscover our earliest and most familiar way of viewing the world. What one calls this 'mutual visibility' is quite irrelevant, be it 'animism' or 'anthropomorphism'. It probably has its roots in our tendency to attribute to the world that it is 'looking out for us' (*auf uns schaue*) when it seems to 'favour' us (in the Austrian sense of 'taking care of us'), while appearing to 'have it in for us' (*absehe*) when it makes us feel threatened.

The following experiment – that is not that easy to conduct – underscores how far from being true it is that someone has to be 'eccentric' in order to feel looked at by the world. On the contrary, a human being has to do violence to their sense of immediacy and naturalness (*Unbefangenheit*) to produce the feeling of *not* being looked at. If we attempt to consciously imagine that things *cannot* see us and do so with ruthless and uncompromising insistence, then the outcome affects us in a most peculiar manner. The idea that the chair on which we usually sit, the table we write at every day, yes, that even our mirror has never ever seen us because it does not know us; the thought that the picture hanging on the right above our bed is plunged into eternal night, that it knows nothing about the companion picture to its left, nothing of its own beauty and finally also nothing about us, even though we look at it every day, the thought, in short, that the picture is blind as a stone and that we have to spend all our days surrounded by totally blind things and have to live seeing while being unseen – this idea is so adventurously alienating that it makes one think one can feel 'the air of other planets'. Only with great difficulty we can fend off the feeling of having just conjured up the image of the ghostly living conditions on a darkened other world.

No, it is entirely unwarranted to believe that we cannot acknowledge things as authorities on account of their 'blindness'. We are far too blind towards the blindness of things to sustain such a belief, as long as we stay uninhibited enough to accept this. It is pure theory to believe that humans reason with themselves 'things do not see me – *ergo* it is not necessary to be ashamed in front of them'. The question is whether such an 'ergo' must be presumed at all. If this is the case, it would probably be entirely different and would very likely attest that our mechanism of deduction unfolds in an anti-clockwise direction: 'I am ashamed in front of something which is superior – *ergo* this thing can see me'.[37]

The components of the objection are hence highly questionable. Rather than pursuing this question any further, it is more important to examine if 'Promethean shame' fulfils the condition that we have established as being the essence of shame. We need to check if it reveals itself as a '*disturbance in processes of identification*', as a faltering process of identification between the 'self and the it' or the 'it and the self'.

It was not without a good reason that the term 'it' was left utterly unqualified when it was introduced earlier. We limited ourselves to using the term to describe everything the 'self' is *nolens volens* '*party-to*' (as it is forced to realise in the disconcerting encounter with itself). The 'it' hence names *what* the 'self' *is also*, without this 'it' being a *part* of the self. It was not permissible to define the concept more narrowly because the most diverse entities emerged as such an 'it': the body, sex and species, '*what one does*' (‘*das man*’).* We were able to content ourselves with not defining the ‘it’ further because we were exclusively interested in the disturbance of identification that sets in when the ‘self’ encounters itself as an ‘it’, that is, we were interested in its character as an ‘ontological endowment’. For, we recognised that the essence of shame is disturbance or confusion of the ‘self’. The phenomena the ‘it’ covers are divergent to such a degree that we do not wish to insinuate that the ‘it’ corresponds to something ontologically homogeneous. To do so would be meaningless.

We cannot be surprised, therefore, that the ‘it’ that we are discussing now is fundamentally different from the quasi ‘natural’ ‘it’ phenomena (the body, sex and genus, the family, etc.) that were previously foregrounded in our analysis of the basic workings of shame. The whole point of our analysis is in fact rooted in this fundamental difference. In ‘Promethean shame’, it is the machine or technological device that is encountered as the ‘it’. We are confronted with this ‘it’ while operating machines, a relationship in which humans have to function ‘with’ and ‘alongside’ technological objects, ‘as’ a part of the working machine and as such they no longer encounter themselves as a ‘self’. To demarcate the ‘it’ that manifests in engagement with machines from the ‘it’ of our previous discussion, we will call it the ‘*it of the technological device* (das Apparat-Es)’.

We can, in a sense, picture the human being – and this is now truly only an image – as clamped between two brackets, as if constrained by two forces that both challenge the ‘self’: on the one side the human is constrained by the ‘natural it’ (by the body, sex and species, and so forth) and on the other side by the ‘artificial’ (bureaucratic and technological) ‘it of the technological device’. The space left open for the ‘self’ is already very narrow today; and because the ‘it of the technological device’ encroaches on this space ever further, this space is getting smaller, while the danger that the ‘self’ is crushed by these two colossal non-individual forces is daily increasing. With every day that passes hope seems more justified – yes hope, because millions hope that this catastrophe will happen and that a technological totalitarianism will set in. When this outcome is finally achieved – tomorrow or the day after – the final triumph will belong exclusively to the machine. In its greed to devour everything and especially that which is alien to it, the machine will not only have incorporated the ‘self’, but also the other ‘it’: the body.

Today technological devices are already trying their best in this respect. The saying that the machine is ‘getting at us and our bodies’ is merely an

* In mentioning ‘*das man*’ the text seems to be making a reference to Heidegger’s *Being and Time*, in which the term is deployed in Section IV (especially, §27) to name the ‘faceless public’ in view of which ‘one just does what one does’. *Das man* refers to the impersonal, indefinite and habitual mode of existence that Heidegger presents as our most basic way of being – Trans.

image, for actually it is 'getting into us'. It even attempts to overpower our sexuality in order to include it in its domain. Here is an example.

14

The orgy of identification as a template of the interference and disturbance in processes of identification – Jazz as an Industrial Dionysus Cult.

Jazz music, which is still today often called 'black' does not owe its existence exclusively – if at all – to 'visceral memories of desert and jungle drum'; it is (at least equally) 'machine-music', a type of music, therefore, which initiates forms of dance that are fitted to the humans of the industrial revolution.* It is not only the 'torpor of an archaic existence' or the 'lust of urgent sexuality' that has become a voice in Jazz, but always also the tenacity of the accurately working machinery of the steel-mill, which impassively and neatly cuts the glissando of animality into ever uniform pieces.

It might appear at first as if the two adversarial, impersonal forces – the 'foundational force of the body, sex and species' and 'the force unleashed by the thing' – are unified in an exhilarated ridicule of the 'self'. It seems as if these two forces flay and whip each other up and work together in mutually reinforcing but yet counterpoised motions to grind up the 'self' caught between a mechanical system that has become orgiastic and an organism that has become mechanical. But in this alliance the machine has not yet achieved its ultimate goal: the machine aims to liquidate the sexed body (*das Geschlecht*)† as such. The alliance with the sexual is not entered into so that the machine can work in unison with it, but in order to transform the trapped sexual energy of the body into energy of its own kind: the machine is attempting to turn the dance into a *process of transformation* and the dancers into transformers whose duty it is to convert animalic into mechanical energy.

Of course we do not mean to say that the vital energy generated while dancing has use-value for machines. But despite this, the machine – in its autocratic aspirations and because it does not tolerate forms of energy from which it cannot benefit – releases the superfluous vital energy from the frenzy of dancing and so exhausts it. And this is done in such a way that *the exhaustion of energy as such takes on the form of machine-like movements*. The types of music which mobilise such forms of dance contain all the hallmarks of an automatic and thing-like motion: their *fury of repetition*, neutralising

* 'Jazz' may here be serving as a blanket description for popular music, but the text here replicates similar cultural prejudices of a number of émigré German thinkers – Trans.

† '*das Geschlecht*' can refer to sex and gender as well has the human species as a whole – Trans.

any sense of time and shattering it, is the fury of the constant metre of the working machine.[38] *Syncopation*, which is the underlying musical principle of Jazz, should not be understood as a purely 'musical phenomenon' of placing a rhythmic accent 'off-beat'. It should rather be seen as the symbol of the ceaseless and never slackening persistence with which the machine cuts in to interrupt the rhythm of the body. In a sense, then, *Syncopation* is the interference with which the machine pounds its victory into every single beat of the music, and the music does so as the continual *refutation of the body* and of its birth-right of 'counting also' and with its own metre.

What the dancer dances is not only the apotheosis of the machine. Because the body co-executes its own refutation in order to prove its conformity with the machine, the dance is simultaneously also a ceremony of abdication and synchronisation. It is *an enthusiastic pantomime with which the body re-enacts its own utter defeat.*

The expression 'Industrial Religion' that we coined above in the discussion of 'Human Engineering' is corroborated here: the orgies these dances portray, for example, in the dance halls of Harlem, have nothing to do with 'pleasure'. They are much more and much less than this: they are *ecstatic sacrificial dances that are cultically celebrated in honour of the Baal of the machine.*

The ecstasy is real. The dancers are genuinely 'beside themselves', not, however, in order to be one with chthonian forces,* but rather to become one with the God of the machine: an *Industrial Dionysus-Cult.* What holds true of syncopation also applies to the 'final cadences' of the music: the victory of the machine-god is drubbed into the body. Expressed in musical terminology, these cadences are always 'false cadences' (*Trugschlüsse*) and like syncopation these are mechanical. They function as '*brakes*' and they represent pure '*deceleration*'. If there were a typology of what meanings 'final cadences' can be made to signify – in late romantic music, for example, the 'final cadence' would signify a 'peaceful and gentle dying off' or 'redemption' – then this 'brake ending' would be absolutely new. Such 'brakes' are '*interruptiones*'. These endings are like *coitus interruptus* (because machines are always stopped at an arbitrary point of time 'when they are right at it'). This expression explains what this really means for the human dancers who have matched their bodily motions to the music. But not clearly enough, for it is not just sexual arousal that is being interrupted, but life as such. When music and dance are slowed down or abruptly switched off like a machine, then the dancer is 'beaten to death' as a whole.[39] He or she seems to have become a victim of murder or an industrial accident. In brief, *when seen as a cult, the interruption signifies sudden violent death.* It is, however, precisely

* *Chthonian forces* are the forces that bring life and end it (fertility and death). These forces are embodied by a number of ancient Greek deities and mythological beings – Trans.

this death that is a central component of the cult, since there is nothing that is more machine-like than being arbitrarily shut down. Being able to end in this way is proof that the transformation into a machine has been accomplished. '*I am being switched off*' – the one who is suddenly being put out of service says and feels – '*therefore I must be part of a machine*'. Being stopped in this way is even enjoyed. One can be certain that the music machine will immediately recommence from the beginning for a new shift, which promises new life in the form of a renewed defeat and a new hope of being murdered again. As such, the interruption represents the culmination of the union with the new deity.

That the dancers are actually 'finished off' by this rite and fully forfeit their sense of self can be seen in a striking phenomenon: during the orgy the dancers *lose their face*. What I mean by this is not the often observed fact that even today's faces are shaped by identical role models and become like similar mass products. Such serially produced faces only differ from one another in the manner in which one piece of cloth is different from another with an identical pattern: they can only be told apart by individual flaws in the weave. No, what I mean to say is that even this mass-produced face is lost. This loss can manifest itself in several ways. The face, for example, can become slack and unkempt in the course of the orgy. It may cease to be the centre of the human being's field of expression and the mirror of individuality. The face *either* mutates into a mere *body part*, the naked and uncontrolled appearance of which no longer surpasses the likeness of a shoulder or backside; *or*, it turns into a blank poker face that is *cut off from the orgy* and so remains in the dark about what is going on downstairs. Here, the face is merely 'carried along' as an 'ontic endowment' because it cannot be handed in at the cloakroom like a jacket before the onset of the ritual. The face can also *glaze over* during the orgy, by which I mean that it clearly ceases to see anything or to be conscious of being visible. It would not be at all surprising, if during the intoxication of his machinisation one of the dancers had the idea of pulling something over his head in order to conceal his face (which in any case no longer has any significance) – for these orgies are not without humour. It would be equally unsurprising, if a new variety of shame were to emerge during such orgies: *facial shame*. I do not mean the shame of having a repulsive or unattractive face, but rather the *shame* the dancer might feel *about having a face in the first place*. Analogous to the ascetic shame about having a body discussed above, facial shame is ashamed of the fact of being condemned to carry around this stigma of selfhood as a compulsory ontic endowment.

Even though these thoughts may be 'philosophical exaggerations' – overstated portrayals (*Entstellungen*) pointing in the direction of the truth – it cannot be denied that the face has become a residue and an obsolete fragment.

It is certainly no coincidence that this loss of the face began in the epoch in which the face was also neglected as a subject for the visual arts and in which a commercial drawing counts as chic when the face is omitted all together in the depiction of humans.[40]

But let's once more turn to the 'machine of the music' that sets this orgy in motion.

When bandleaders attempt to perform this music in the form of 'serious music', that is as a concert in which the music should merely be listened to, they are falling victim to a misunderstanding. This, in turn, stems from an appetite to provide the music with the social respectability of acknowledged 'cultural values'. This aspiration is not misguided on account of the music being too 'light' or merely being 'pop'. The opposite is the case. The music is terribly serious, too serious for the concert hall. What I mean is that it affects humans too deeply and violently and in a manner that alters their '*ethos*' (in the musical sense of the ancient Greeks)* much more radically than solemn concert music can today. Once it has reached its final symphonic apotheosis, classical music leaves the listeners – who remain a mere audience – with an 'experience' that evaporates straight away, because it has nothing to do with life outside the concert hall. There is nothing more inconsequential than the effect of serious music. There is nothing more serious, on the other hand, that is, nothing more dangerous or more destructive than the effect of the type of music so often dismissed as being 'superficial'. This music constitutes a real intervention, a violent metamorphosis that has an awful lot to do with the world and life outside of the concert hall. The mode of being it works its listeners up to is that of the machine, a mode of being, therefore, which rules from morning to evening and the music now makes people conform totally with this mode of being.

Because every ritual demands *full bodily participation* this is also required here. That is also why those who stage this music '*in concerto*' merely as art, in front of a mere audience, do not create audiences that are newly receptive for art or self-improvement; rather, they betray the cultic-community. Its members, instead of being able to participate in the mysteries of the cult, are condemned to content themselves with the role of mere listeners. They have to be mere voyeurs of machinisation, so to speak. That such audiences, unwilling to be fobbed of with this role, begin to revolt in outrage about the fraud, as was recently the case in Vienna during a Jazz performance 'in concert', is truly anything but surprising.

What we have just portrayed is a situation in which an ecstatic rite acts as a means of violence to force an identification with the machine. Because no

* In ancient Greek the word *ethos* could be used to describe the formative power of music and its ability to structure thoughts, feelings and attitudes – Trans.

real machine surfaced in our discussion, the situation described is in a sense delusional – it represents a 'pleasant' or macabre 'illusion'. Unless of course one views the music itself as a 'machine' that was created by humans as the special apparatus with which they can accomplish their own machinisation.

To explain why this excursus is connected to our main investigation is hardly necessary. It is obvious that a separate and specific initiation rite never would have been invented, never would have needed to be invented, if humans were usually able to identify successfully with the world of appliances and machines. They would not have needed such an orgy to savour this identity. The fact that the rite exists can thus be seen as a symptom, or at least as an indicator of a symptom, that such identification usually fails.

But we do not yet want to accept this indication as constituting proof. We will rather 'test' it further by scrutinising the 'natural' relationship between humans and machines. That is, the relationship that realises itself in the workplace and examine whether it involves a confrontation of the self with itself in which disturbances in processes of identification become acute. And finally – and with this we return to our original question – we will ask if this interference is 'shame'.

15

The Relapse – The self encounters itself as a failing self – The example of failing labour as the prime test-case of 'Promethean shame'.

What can it mean to have a 'disruption of identity' (*Identitätsstörung*) in the face of machines?

In an analogy with the twofold principle outlined above – the 'I is ashamed of the it' and 'the it is ashamed of the I' – this question can once again be answered in two ways:

Either: instead of encountering themselves when using machines, humans find something that already conforms to the world of machines: they discover themselves as part of a machine.

Or: humans have already willingly integrated themselves with a machine (or the world of machines as a whole). Despite this, because they are unable to 'loyally hold the line' and achieve a full conversion and co-substantiality with machines, they now encounter themselves instead of finding a machine part.

In the first case the 'I' encounters itself as an 'it'; in the latter instance the 'it' meets itself as an 'I'.

Discussing the first of these possibilities any further is unnecessary, because it is already obsolete. The type of human being that *Chaplin* portrayed in *Modern Times* who compulsively enacts involuntary, machine-like movements and continues to do so even when he is no longer operating a machine does not exist. There is no human being who realises with feelings of alienation and horror that he has already become a part of a machine. No, such Chaplinesque humans do not exist. The portrayal is distorted. It is the other way round. 'Modern man' is disconcerted, rather, because a resilient residue of self still remains even when he 'actually' ought to be nothing – and 'actually' wants to be nothing – other than a 'cog' in a machine. What shocks him, for example, is that he cannot keep up with the conveyor belt he is standing at because it is moving too fast or because his body cannot adjust itself to the combinations of movements it is asked to perform to work with the machine. Alternatively, he may be lost in thought or he may simply desperately want to scratch himself because of an itch. Nothing as scandalous as this can happen to any other part of a machine.

But let's begin at the beginning, with the process, therefore, by which workers should become familiar with the way the machine works. Let's see if this too already leads to something like an ambiguity in the self-identity of workers.

Anyone who has been confronted with a new job at a conveyor belt knows what effort it takes to transform this first confrontation into a synchronisation to the working rhythm of the machine. They know how hard it is to keep up with a working machine and are familiar with the fear of not being able to keep up. Because it is mundane, the antimony that this initial training causes usually remains invisible. If one considers, however, that in order to be able to work *without difficulty*, workers have *to exert* themselves in a most concentrated manner in order to take on the pace and rhythm of the machine; if one contemplates that workers are required to *exercise the strictest self-control* in order to start *an automatism*, that they have to *pull themselves together* so as to not *function as themselves*, then one has to admit that this task is paradoxical. The ordinary expressions 'adaptation' and 'training' only provide names for this exercise. The paradoxical demands that these processes make on workers – these are asked to extinguish what they are *as* acting individuals, and *transform* their actions into *mere processes*, heteronomous and automatic ones at that, and, once this automatism has been acquired through successful transformation, to scrupulously control and uphold this automatism – these contradictions are utterly concealed by the expressions 'adaptation' and 'training'. The suggestion, moreover, that all doing in accordance with an implement requires a level of 'adaptation', whether using a hammer or playing the violin, and that such 'adaptation' cannot therefore be considered to be degrading to human dignity, does not offer any evidence against the paradoxical nature of this task. For sure, the violin player has to become attuned to their instrument. He or she has to adjust the strokes of their bow as the instrument and the movement of the music require, and must even make sure that this becomes 'natural and intuitive'. In comparison with the induction of factory workers, however, this task is still perfectly human. It is free from contradiction in as much as the violin player is allowed to be unambiguously active while learning to master the instrument, thus transforming the violin into a part of their body (which increases its field of expression), incorporating it as a new organ into the organism. The labour of adaptation that workers are asked to perform at the workplace, on the other hand, is the absolute inversion of this task. They are asked to turn themselves into an organ of the machine. Their task is to *let themselves be ingested* into the movement of the machine; they must put themselves in a position in which they are *incorporated* by the processes of the machine. In brief, *they are asked to actively take into their own hands and carry out the process by which they become passive machine parts.* The paradoxical nature of this expectation is indisputable.

In other words: Because the workers are required to mobilise all their powers of concentration to try to displace their focus away from themselves onto the machine, they must *simultaneously* be *'themselves'* and not be *'themselves'*. We are already familiar with this formula. We employed it above to describe the ambiguous identity of the person experiencing shame. It is of course no coincidence that it already appears here in the description of the worker's first familiarisation with the workings of a machine. In fact, we are already in the close vicinity of shame, our problematic. We will reach its core when we give an account for the self-encounter in which this ambiguity becomes acute.

Answering this question is only difficult if we believe that we have to look for an explicitly theoretical act of reflection. It would certainly be pointless to try and locate such an act. The worker's *encounter with him- or herself happens in the course of using and operating the machine.* It is a moment of work itself. Or, if one understands the expression 'moment' as a reference to time, this encounter is a particular instant in the labour process.

Let's assume that a worker has become successfully attuned to the machine. As long as the work progresses smoothly and without friction between human and machine, the worker is not 'self-aware'. In fact, as long as the worker co-functions according to the rules and as a 'convert' who has become a 'cog' in the machine, his or her self is 'not present' and has no being at all, at least not as a self.[41] Workers only 'assume a self' when their conformity is insufficient, or when their work suddenly goes wrong. It is only then that the self 'becomes conscious' and encounters itself, namely as something objectionable: as a failure.

This means that the self only attracts its own attention because it falls out (of the 'it of the technological device' and its conformist existence with the machine), because it has become visible as a counterforce, as a resistant part that works against the machine. The contours of the worker's individuality only become visible because – to use the words of a famous formula – it is a '*negation*' (an anti-thesis – Trans). Put even more clearly, *the encounter with the self is not the cause of a 'disturbance of identity'* (Identitätsstörung)*; it is the other way round. This encounter only occurs because there is interference* (Störung).*

But *who* now encounters *whom* in this case?

The answer certainly is: the self encounters the self. But is this response adequate?

It would be totally wrong if this were taken to mean that 'the self that is affected' is an equal partner or even identical with 'the self that mobilises this affect'.

'The self that mobilises this affect' is the exemplar, the 'it-and-thing-like' machine-self, the co-functioning and conformist self that has been transformed into a part of the machine and so has already left being a self behind. At least, it had hoped to have left this behind. What surfaces as 'the affected self', on the other hand, is the old self, the old-fashioned being, lagging behind, that might have once had a *raison d'être* in ancient times before the birth of the machine. Now, however, at a time when adaptation to the 'it of the technological device' has become obligatory, this old self has lost its right to exist.

Or if one prefers not to speak of two selves in this manner, but would rather see this as two different states of the self: the encounter with the self only sets in because the conformist self '*gives way*' (*nachlässt*) and so '*relapses*' into its old condition; for example because it dwells on its thoughts, or has a headache and therefore forfeits its previously achieved state of perfect conformity. If the expressions 'giving away and relapsing' (*Nachlässigkeit* and *Rückfälligkeit*) sound moralistic,† this is unintended. The association of 'encountering the self' and 'to fall' with which we are familiar (e.g., from phenomena such as having a 'guilty conscience') is a fact with the widest possible validity. And if it appears here, this is not because we are trying to force an analogy with moral shame, but simply because it too becomes visible while observing the present situation.

This portrayal of the one 'affected', however, is still incomplete. We said that it is the 'old self' that is 'affected' in the encounter. Is it really only this?

---

* The term *Identitätsstörung*, which I have translated as 'disturbance of identity' throughout, here takes on an additional nuance. The text begins to foreground the word '*Störung*' which has several specifically machine related meanings, such as, interference (in a radio or phone signal), a malfunction or a defect. I have used 'interference' here, because Anders's writings frequently evoke the phenomenon of being uncomfortably confronted with oneself on account of a power cut or a missing signal, or as in this instance, when a human factor interferes with the working of a machine – Trans.

† *Nachlässigkeit* also means negligence – Trans.

Let's once again assume that the self encounters itself as a failure. Measured by the intransigent yardstick of the machine, it is of course entirely inconsequential *who* disrupted its function, be it the self or its body. The body and the self are both equally incongruent with the machine; in fact, for the machine this difference does not exist at all. Figuratively speaking, the machine would shrug its shoulders, if a worker had the idea of justifying his failure by blaming his clumsy hands, which he can do nothing about.

But the culprit does not seek excuses of this kind at all. For if in the instant of failure he is forced to rediscover himself as the old Adam rather than as part of the machine, he judges himself by *its* measure, views himself from *its* perspective. It is as if he were a loyal party member who declares, yes even feels himself guilty in the interests of the party, which must be defended at all costs against any deviation from the party line. Even for him the distinction between body and self – of these old antagonists who once played such a fundamental role (because the self was ashamed of the body) – has lost all interest. In fact, the difference has become so irrelevant that the two seem to represent one unified formation. Instead of the rift between the body on one side and the self on the other there is now only the rift between the machine (or rather between the conformist self which is representative of the machine) and the old residue. This residue is made up of body and self in an undifferentiated manner.

These are thus the partners in this encounter, and these the antagonists.

But how do they relate to one another?

With this question we have reached the critical point of our investigation. We no longer deem it possible to describe this encounter with any other word than 'shame' (in the proper sense of the word).

If the example with which we opened our investigation – T.'s attendance at an industrial exhibition – still invited the suspicion that we are only using the expression 'shame' metaphorically, this was understandable. For the situation of merely looking at technological devices is not the one in which shame actually erupts. The perfection of the product is here only theoretically compared with the imperfection of the body.

But now we have left this situation behind and have replaced it with one that the appliances, devices and machines demand from us: the *operation of machines*.* If it were merely a question of certain 'psychological difficulties' that the observation of machines can lead to, as it was in T.'s case, then

* The word *Gerätbedienung*, translated here as 'operation of machines', could also have been translated literally as 'serving the machines'. Anders elaborates on the word *Bedienen* in the *The Obsolescence of Human Beings Vol. 2*: 'If we were a bit more keen-eared we would probably be astonished every day to realise that the word *Bedienung* that once was employed to describe what servants "do" when serving "ladies and gentlemen" […] now describes what we do when operating machines' (p. 69) – Trans.

we could simply go back to our daily business and dismiss this difficulty as an emotional luxury. But the difficulty of operating a machine, the efforts required in learning how to do so, the constant threat of failure and finally the actual failure are essential aspects of operating a machine and emotional processes that are not just metaphorical. The effect of this actual failure is not limited to highlighting the different levels of perfection between humans and machines, that is, between the producer and the product. The failure rather 'ejects' (*wirft heraus*) the one who fails. Throwing him or her back onto themself so that their old residue of self now suddenly stands there without a world, inept, 'rejected and disproven' (*verworfen*), no longer knowing what to do with itself.*

It is simply not possible for them to identify with the 'unadaptable fellow', the reject one he or she finds themself to be. It is impossible to be content with that 'residue standing there'. *As* the conformist who they have been up to this moment, and who they actually still are, he or she 'is' not a 'residue' and 'is' neither this body nor this self. At the same time, however, it is impossible for them *not* to identify with this residue: for they cannot deny that with which they have been 'ontologically endowed'. And as we know from earlier, the fact that this endowment was not their fault does not offer any absolution, rather, it precisely occasions shame and disgrace: for precisely the fate that they cannot do anything about is 'fatal'. And so they are ashamed.

As difficult and complicated as this may sound in the language of theory, this does of course not put an end to the intensity of the actual condition. The one who is ashamed has no idea about theoretical difficulties. He or she knows nothing at all. For, to be ashamed means to be ignorant. To be ignorant, in turn, means to be completely at a loss with nowhere to turn. And this state of being at a loss is now precisely his or her condition. Having fallen out of the state of 'being it', a state to which they had still inconspicuously belonged just a moment before and which they recognised as exemplary until they were jettisoned, they now stand publicly exposed, in full view of that which they have squandered. Instead, they have to be a 'self' which is nothing more than a 'deficient mode' of being a machine, nothing other than a scandalous non-device and a conspicuous nobody – despite having a definite name, a definite body, or maybe even portraying the definite flaw of an individual peculiarity. And although they no longer have any business in the realm in which the machine reigns, they still have to stay there, for not to be there also exceeds their powers. Thus they have no other wish than to

* The text is here playing on variations of the German word *werfen* (to throw) in a manner that suggests an engagement with Heidegger's concept of *Geworfenheit* ('throwness'), a concept, which Anders has already evoked above in endnote number 2 – Trans.

sink into the ground, nothing else than this utterly unaccomplishable desire, only shame.

Anyone who has never blundered while handling a machine; anyone who has never incredulously looked back in disbelief at the assembly line that wordlessly keeps on moving; anyone who has never found themself cast onto the shore of their old self and never experienced what it means to suddenly find there *oneself*, precisely oneself; anyone who has never looked down at their hands in an alienated and bemused manner, at these clumsy instruments here, because their obsolescence and incorrigible incompetence have caused the fall from grace – this person does not know the shame that is the shame of today, the shame that erupts daily a thousand times over. Anyone who denies the existence of this form of shame does so only because to admit that we have come such a gloriously long way to arrive at the ability to be ashamed in front of things makes them blush with shame.

NOTES

1. 'To the Cogs', in *Molussic Industrial Hymns*, translated into German by G. A. [Molussia is a fictitious land of Anders's invention. Anders's writings frequently evoke stories, sayings and songs from Molussia as a stylistic device. The short story 'der Ahnenmord' even contains an ethnological account of Molussia that makes reference to fictitious academic events and works with English titles, such as *Molussic Excavations, the Tablets A-D. Preliminary report by Sir Allen J. Godefroy*, Calcutta University Press 1932, *Molussic Studies*, a Symposium, Princeton 1952 and *Molussic Proverbs*, London 1949. See: Günther Anders, 'Der Ahnenmord', *Die Kosmologische Humoreske und andere Erzählungen* (Frankfurt: Suhrkamp, 1978), 241–95. The substantial antifascist novel *Die molussische Katakombe* (*The Molussian Catacomb*), which Anders wrote between 1930 and 1932, seems to be the earliest, archetypical gesture of finding recourse to the fiction of Molussia as a means to critique the present. The political situation in Germany made the publication of the novel impossible at the time and it was only posthumously published in 1992. – Trans.]

2. The rejection of being born (*natum esse*) is a motif that constantly reoccurs, especially in the context of religion. Founding figures are often stripped of the flaw of having been born (Moses). The last echo of this religious rejection can be found in reactions to theories of natural descent, that is, in the indignation with which the proposition has been met that we humans have evolved (from other entities). The latest variant of such discreditation of the fact of birth comes from the philosophy that accompanied the bourgeois revolution. *Fichte's* notion that the 'I posits itself' is the speculative version of the self-made man, who does not want to be born or have become, but rather wishes to owe his existence to himself as the product of his own making. This particular rejection of the fact of birth has its origins in the resistance to the notion that a 'noble' or 'lowly birth' is the source of either privilege or the lack of rights. In Fichte's philosophy the wish to 'be made' (i.e., to be self-made)

rather than being born did of course not have any technological connotations, but was exclusively moral and political: the self-made man is autonomous and the citizen of a state which is made by citizens. *Fichte's* famous omission of a philosophy of nature has its origin in this resistance to *natum esse*: As *Schelling* writes, 'The arrogance of man rises up (*sträubt sich*) against this origin from the ground' [F. W. J. Schelling, *Philosophical Investigations into the Nature of Human Freedom*, trans. Jeff Love and Johannes Schmidt (Albany: State University of New York Press, 2006), 29]. *Heidegger* is a later variant of Fichte. His notion of the '*thrownness of Dasein*' is a protest not only against being created by God, that is, against supernatural origins, but also against having developed, that is, against natural origins. For further elaboration on this point see the author's 'On the Pseudo-Concreteness of Heidegger's Philosophy', *Philosophy and Phenomenological Research* 8.3 (March 1947), 337–71. [Anders published this article in English – Trans.]. 'Thrownness' is an *evasive-idea*. With Heidegger, however, the process of self-fashioning (the transformation of 'Dasein' into 'existence') has lost all its political connotations. The process remains a purely solipsistic enterprise. It is no coincidence that (at least at the time of *Being and Time*) natural philosophy is missing as in the case of Fichte: If he had not left it standing at the door, his denial of the fact of birth could not have been accomplished.

3. 'To the Cogs', in *Molussic Industrial Hymns*, translated into German by G. A.

4. On this point, see the more detailed analysis of shame in Section 11 below.

5. The desertion of the one who feels threatened into the camp, or at least into the value-system, of the one who extends the threat is a motif we are well acquainted with in the context of today's politics. And in the last two and a half decades [1930s to the mid-1950s – Trans.] we have often enough witnessed how those who are threatened not only take on the criteria of judgement of those who extend the threat, but also their feelings. They are almost always brought to the point where they do this in a craze of voluntariness. In fact, this more or less gentle enforcement of voluntary conduct is the counter-revolutionary accomplishment of our times. Outside of the political arena, however, this process has as yet not been documented.

6. Nothing is more misleading than to subsume this 'overcoming of the body' under the category 'jewellery'. It is nothing less than the inversion of this category. Whereas jewellery seeks to enhance the beauty of the living body, bodily modification aims to lend the body the beauty of fabricated things.

7. See Chapter II ['Reaching out for the Sky' – Trans.] of Robert Jungk's *Die Zukunft hat schon begonnen* (*The Future Has Already Begun*). This example is in keeping with an account provided by *Thomas Power*, the commander of the American Air Research and Development Command: In May 1956 an H-bomb had missed its target by 6 kilometres. When called upon to account for a mistake of such gravity, Power, according to Reuters, offered the following explanation: 'when dealing with human beings such things can just happen'. The pilot had forgotten to operate a certain lever. One ought to listen to Power's words carefully. Anyone who only hears the old saying 'to err is human' misunderstands their meaning. They imply much more: Because it is human to err, that is, because the human operates so unreliably, it is actually inappropriate to entrust humans with such a perfect machine. The human is

here seen primarily as a *source of error* [See: Robert Jungk, *Die Zukunft hat schon begonnen* (Stuttgart: Scherz und Goverts, 1952). – Trans.]

8. The failure is 'shameful'. As we will see below, the fact that a failure is not one's fault does not make a defect less shameful, but more so. See Ch1, 65ff*.

9. In his 1930 essay '*Die Weltfremdheit des Menschens*' ('The Human's Estrangement from the World'), which appeared in French translation as '*Pathologie de la Liberté*' in the *Recherches Philosophiques* in 1936, the present author defined human beings as 'indeterminate', as '*indéfini*' and 'unfinished' – as '*free and indefinable beings*' who are only defined and definable by what they make of themselves in each specific instance. (And soon after, Jean-Paul Sartre also formulated his credo along very similar lines.) By over-exaggerating the philosophic-anthropological dimension of freedom, both Sartre's and my own early work represent belated attempts to obscure the then already valid circumstance that the 'subjects of freedom and unfreedom' need to be substituted. Such definitions only seem plausible because animal existence as a whole is here used as a foil against which the human is measured (and this, by the way, is the case in almost all non-theological anthropologies). They seem plausible, therefore, only because '*the* animal' (an expression which already represents an ad hoc abstraction) is presumed to be unfree, a prisoner of the fate of its species. The trouble was not taken to verify this presupposition, because it was considered self-evident (not least because of the theological tradition). Today, I find this highly questionable, not least because I deem it philosophically audacious to use a foil for the definition of the human that does not match up to the effective backdrop of human existence. After all, we do not live against the backdrop of bees, crabs or chimpanzees, but rather light bulb factories and radio sets. But even from the perspective of natural philosophy, the opposition between 'man and animal' is unacceptable. The idea that the single species 'human' can be opposed to the thousands upon thousands of immensely different animal species and types is simply an anthropocentric delusion of grandeur. As a warning against this cosmic immodesty, each textbook of 'Philosophical Anthropology' should have '*the Fable of the Ants*' as a preface. This fable recounts how, at their universities, ants learn to distinguish between 'plants, animals, and *ants*'. If one looks at what actually forms the backdrop to human existence, that is, the man-made world of products instead of choosing 'the animal kingdom', then the image of 'the human' changes immediately. The singular word 'the' human immediately disintegrates, and with it, the notion of human freedom. [The above-mentioned significant essay '*Pathologie de la Liberté*' is available in English translation. See: Günther Stern (Anders), 'The Pathology of Freedom: An Essay on Non-Identification', trans. Katharine Wolfe, *Deleuze Studies* 3 (2009), 278–310. (All future references are to this edition). A revised version of this translation can also be found in: Günther Bischof, Jason Dawsey and Bernhard Fetz (eds.), *The Life and Work of Günther Anders* (Innsbruck: Studien Verlag, 2014), 145–70. – Trans.]

10. By the way: a certain boomerang effect is part of every machine in so far as it is sent out to do its job and 'bring in' a return.

11. 'To the Cogs', in *Molussic Industrial Hymns*, translated into German by G. A.

12. This term was created ad hoc following the template of Karl Jaspers's famous discussion [of 'limit situations'. The notion of 'limit situations' is a central component

of Jaspers's Existentialism. They describe confrontations with limits that cannot be transcended, altered or surpassed because they fundamentally demarcate finite human existence (examples of such limits are chance, death, suffering and guilt). See Karl Jaspers, 'Chapter 7: Limit Situations', *Philosophy Vol. 2: The Illumination of Existence*, trans. E. B. Ashton (Chicago: Chicago University Press, 1969) – Trans.].

13. The excessively strenuous demands which humans here place on their bodies so they can live up to the excessive tasks of their machines resemble, in a highly peculiar manner, the excessive demands which those who practice speculative metaphysics have placed on reason: both *ignore* the *fact that human power is limited* (*Faktum der Leistungsgrenze*); here too the limit should be extended or destroyed. Only that the human in this instance does not pretend to be omniscient '*sicut deus*' (like god), but aims at becoming '*sicut gadget*' (like a gadget).

14. Today every industry produces these two kinds of product. The branch of production which makes the second product (the product which generates the need for the first product) is called the 'Publicity Department'.

15. I say 'robots'; and I say 'conformity' (*Gleichschaltung*). 'Robots' because the true robots of today are not the pretend-humans (*Scheinmenschen*) that are assembled from dead components and actualised in 'computing machines', but the machine-parts that living human beings function as. And I say 'conformity' because the process in question here is a variant of the same behaviour we are familiar with in the context of political reality as the 'enforced conformity that totalitarian systems of government effect'. Only that the variant treated here is more pronounced because humans are in this case actually attempting to 'turn themselves into things', whereas in the political sphere enforced conformity through reification only ever turns people into a 'thing' metaphorically (even if these people have been utterly robbed of their freedom).

16. While employed as a worker in California, the present author had to listen to the phrase 'forget about it' more often than to any other. In this impatient reminder not to know more than one is required to know, a resentment of the 'highbrow' unfailingly resonates. It is directed at anyone who has openly provided proof in the form of too much knowledge or too many interests that they are (or have pretensions to something) 'better' and that they do not belong to the people.

17. [The text is here replicating a formula that is frequently deployed in Anders's writings, usually with overtly moral overtones. The poem 'Humans, you are smaller than you are', which was published on the 8th of August 1940 in the New York based German Jewish newspaper *Der Aufbau*, is exemplary in this respect. It is worth replicating the poem in translation here: 'Sure, you can kill a thousand / but your tears can mourn no more than one / How small sadness's compass is / Poor, abjectly meagre we wished and made the soul / a fool, who called it godly, infinite / See, the power given you to act is greater / more tremendous than the space freed in your heart to comprehend / Your heart is outclassed by your hand's blind grip / This instructs / Humans, you are smaller than you are.' The idea that humans are smaller than they are introduces the central problematic that Anders philosophy of technology seeks to think, namely that technological advancement leads to 'a Promethean discrepancy', a growing rift between 'making (*Machen*) and imagining (*Vorstellen*); between doing (*Tun*) and feeling (*Fühlen*); between *knowledge* (*Wissen*) and *conscience* (*Gewissen*)' – Trans.]

18. I don't call those 'metaphysically conservative' who insist on referring to existing metaphysical philosophies. I am rather talking about those, who (in parallel with the politically conservative) insist on a (metaphysical) state that they assume the world to have.

19. In the final instance, the 'autarchy of the moral' as well as 'nihilism' are both just euphemisms, with different emphases, for the way in which the moral has this quality of 'being like a metaphysically cut-flower'.

20. By the way, for many of these products 'immortality through replaceability' might already be redundant, because they can be made to be practically indestructible. This is the case, for example, with certain razor blades. The patents for these blades have been bought by companies that seek to prevent their production because the actual immortality of a particular razor blade would result in the death of razor blade production. Production lives from the death of the single product (which has to be purchased again and again). In other words, the 'eternity' of production rests on the mortality of its products.

21. Only the logician finds it hard to bring this ideal of conformity into synch with the ideals of 'freedom' and 'individuality' (for these two ideals have not been thrown overboard). In our real historical situation, they do not interfere with each other, for the unfree are simply *declared to be* 'free'. Because conformists enjoy the liberty to absolve their happy extrovert career paths, to which they have a right, without coming up against the edges of their own individuality, *they* are now those who are 'free'. 'Freedom' is hence equated to 'smoothness' and 'frictionless advance', whereas 'individuality' becomes synonymous with being entangled in 'barbed wire' and with 'disablement' (*Behinderung*). Incidentally, in contrast to theoretical systems, there is no actual social order that cannot afford to sustain contradictory ideas. As long as each of these ideas has a function that is useful for a given order, then nothing stands in the way of their coexistence (e.g., the coexistence of the first book of 'Genesis' and the principles of 'genetics'). On the contrary, only humans who point to these contradictions are deemed oppositional and their impolite fanaticism for the truth makes them suspect of being intolerant. As a matter of fact, there is no contradiction or inconsistency that could not be justified through a false appeal to tolerance.

22. It is therefore plausible that death and dying become more conspicuous when the exemplary and enviable status of products is widely acknowledged because a given country has high industrial potential, and because in this country the frailty of being human plays itself out in front of the backdrop of deathless serial products. Death is here more likely to be regarded as an unnatural embarrassment than in industrially less advanced countries, in which the mortality of creatures has not yet been covered over. It is hence plausible that in industrialised countries the attempt to hush up the fact of death is more widespread and conducted more vehemently than elsewhere. I am, of course, speaking about the United States, whose prevalent attitude towards death will probably be adopted by us all in the future. For it is well known that in the States it is a widespread custom to apply make-up to corpses for their funeral, in order to transform the dead body *post festum* (after the event) into a pretty ready-made product and to undress the corps from its deadness. Furthermore, cemeteries are advertised as 'lovely sites' with the aid of the tempting sentimental images

(in the style we know from travel advertisements). The living, that is, those who will die (the *morituros*), are hence addressed as if they will become 'posthumously immortal'. The suggestion is that they will be able to daily enjoy the pretty setting like pensioners, once they have moved in and taken 'possession of' their graves. It can hardly be denied that, no matter how inarticulately, a *preservation of life after death* is implied here (one, of course, that has nothing to do with religious beliefs). The little book by Evelyn Waugh dedicated to this macabre business of immortality is also well known here. [The reference is to Evelyn Waugh's highly satirical novel *The Loved One: An Anglo-American Tragedy*, first published in 1948. The novel revolves around Dennis Barlow, a British poet who begins working at the pet cemetery and funeral service 'Happier Hunting Ground' after losing his job as a Hollywood scriptwriter. The story follows Barlow's cynical attempts to win over Aimée Thanatogenos, a cosmetician employed as a mortician's assistant at the (human) cemetery 'The Whispering Glades' – Trans.]

23. The fact that the engineers were incapable of associating anything other than 'to feed' when responding to the word 'brain', is to me the icing on the cake.

24. This type of shame is of course not a variant of 'Promethean shame'. On the contrary, for anyone who is ashamed in this new sense, the 'disgrace' is the very existence of 'Promethean shame', the fact that there is such a thing. This is thus 'shame about being ashamed', that is, 'iterated shame'. As such, this variant may be the first corrective remedy for 'Promethean shame'.

25. The philosophical novelist may even be inclined to weave this dialectical thread further into a text that is now entirely fictitious. He would show how MacArthur's act of revenge was always destined to fail, that he had to 'find death in triumph' because he had to learn from experience that the word of machine oracles is as indispensable to the management of companies as it is for the waging of war. That is to say that in the interest of revenge, MacArthur had to make use of the verdicts issued by the very creatures he wanted to revenge himself upon. These machines thus had ample opportunity to take pleasure in ridiculing him and to spoil the satisfaction he sought to obtain. This is not the place to follow the 'dialectic of Promethean shame' all the way to this point.

26. Because the 'authority' in front of which one feels ashamed is often 'oneself' – and this is especially the case in historically recent varieties of shame – shame can be reflexive three times over: *One* is ashamed; one is ashamed *of oneself*; one is ashamed of oneself *before oneself*.

27. This originally 'negatively intentional' shyness is transformed into positive religious awe the very moment it turns to the authority in some way instead of fleeing it. Now it anxiously faces and subjects itself to this authority from a distance. When this happens, shyness turns into 'awe' in the shape of quivering worship (of that which is '*awe-inspiring*'). Because the worshipper before the authority is a 'nobody', the 'interference in identity' sets in which gives shyness the character of 'shame'.

28. I do not believe that there can be a philosophical system based on reflection (*Reflexionsphilosophie*) departing from the fact of the self's recognition of itself (*Faktum des sich selbst vorfindenden Ich*) that is not either a philosophy of freedom or unfreedom, or a philosophy of pride or shame. In philosophies of shame,

identification fails; in philosophies of pride, the 'I' occupies, or 'constitutes' the 'me', or it simply denies its existence.

29. On this point see my 'Pathology of Freedom', 283f.

30. When 'Dasein', at the beginning of *Being and Time*, asks about its 'who?', and finds itself, even though it crystallises into a 'self' in asking this question, as already being constituted by the 'impersonal public' Heidegger calls '*das Man*', then 'Dasein' discovers itself as an 'it'. In fact, the action *Being and Time* portrays (in the form of a theoretical ontology) is nothing other than a systematic fight against shame. *Being and Time* is about the attempt of the 'I' (which is ashamed-of-itself) to overcome the disgrace of being an 'it' and to become 'itself'. [See *Being and Time* Section 9 for Heidegger's introductory discussion of the question 'who?' – Trans.]

31. The conventional assumption that shame occurs especially or even only in the case of something 'that was one's fault' clearly inverts the true state of affairs. This assumption is a symptom of the immoderate pretensions we have about freedom, which we just mentioned, whereby humans attempt to appropriate the pain of unfreedom (which shame represents) in order to present it as a 'pain that acts as punishment' and employ it as such. The truth, rather, is that one is ashamed of an action because in the eyes of the authority one is someone that one should not be. How little it is actual guilt and responsibility that generates shame is evident in the fact that one is also ashamed when wrongly accused, yes, especially then. In this instance one is not ashamed because one is hurtfully believed to be capable of this or that misdeed – this would be far too subtle to be true. No, one is ashamed because in the eyes of others, that is, socially, one actually is guilty. In brief: one is not ashamed of guilt; on the contrary, that of which one is ashamed often turns into guilt. '*I am ashamed – therefore I am guilty*' *is more valid than* '*I am guilty – therefore I am ashamed*'. In dictatorships, for example, it has happened hundreds of times that actions or failures to act became shameful the very moment when they had to be justified in front of a jury, even if these acts or omissions were motivated by moral concerns. And it cannot be denied that the onset of shame not only convinced the jury that the accused was guilty, but also the accused.

32. See objection number 2 as outlined in Section 2.

33. Philosophically speaking, Sigmund Freud's discovery of the 'Trauma of Birth' cannot be rated highly enough: for what more incisive event could happen to life than being torn from the 'ground'? The feelings Freud brought into view ('Oceanic Feeling'; 'Death Drive') are metaphysical in every respect, even if Freud's language was masked by the scientific vocabulary of his century. This is equally true of the 'Trauma of Birth' with which he described the *shock of individuation*, no matter how well he disguised it. In an analogy with our question 'who is ashamed?' the question to ask in relation to this trauma is the following: '*Who* is actually shocked here?' Is it the individual itself? Or, since the shock precedes individuation, is it not rather 'life that has not yet been individuated'? Is not the process of individuation shocking *in itself?* Is the individual not merely the *heir* to this shock, which it now drags with it throughout its life as the pain of being an individual, a pain that is never overcome? It seems to me that this was Freud's view; for the 'Death Drive' is ultimately nothing other than the longing of the individual to be rid of the agony of being an individual.

34. This theory of shame was first formulated in 'The Pathology of Freedom', where I employed the expression 'leap from the origin (*Sprung aus dem Ursprung*)'. [See 'The Pathology of Freedom', 288ff. – Trans.]

35. This type of shame includes what is probably the most basic form of shame – metaphysical shame: the shame the individual feels at having been taken out of the 'ground' and existing *as* an individual, *as* a singular being: the shame of having 'fallen off' the ground (*abgefallen*) instead of still belonging to it in the cosmically proper manner; the shame of now being forced to exist as a '*cosmic exception*'. The equation of this form of shame is: Non-belonging = improperness = insubordination. Anaximander and (Hölderlin's) Empedocles are the principle witnesses of this form of shame.

The following can be read in the Molussian metaphysics of individuation, which did not yet know our speculative inhibitions: 'the ground-of-the-world (*Weltgrund*) is ashamed of its individuations, of the fact, that it has distanced itself from itself in the form of individuals.' Further we can read (in a similar vein to a famous formula by Spinoza): 'the shame the individual suffers from is only a part of the shame with which the "ground" is ashamed of its individuations.' The '*it*'.

36. It is hard to conceive a philosophically more ridiculous preoccupation than the one concerned with the variants of what is called 'synaesthesia' (the allegedly retroactive confluence of characteristic impressions belonging to divergent sensual realms). Those who are astounded by these phenomena should also be astonished that tree branches link up with a tree trunk. The philosophical problem is not the question of how these divergent characteristics 'converge', but rather how these characteristics branch out from their originary, pre-specific state in the first place.

37. After completing my book, I have found that an emphatic account of this mutuality of visibility is part of Walter Benjamin's discussion of his notion of 'aura' in his admirable Baudelaire-Essay. Benjamin's account anticipates what is asserted here. In the same essay there is also a treatment of Valéry's characterisation of perception in dreams: 'In dreams there *is* an equation. The things I look at see me just as much as I see them.' [See Walter Benjamin, 'Some Motifs in Baudelaire', in *Walter Benjamin: Selected Writings, Volume 4: 1938–1940*, ed. Howard Eiland and Michael W. Jennings, trans. Harry Zohn (Cambridge, MA: Belknap Press, 2003), 313–55, 339. – Trans.]

38. The music-machine especially likes to incorporate, with special glee, monotonous folkloric elements to this end.

39. [It is worth noting that Anders, who lived in New York in the late 1940s, is here responding allergically to the music and cultural milieu that gave rise to the first celebratory articulations of 'the Beat Generation'. These see in the energy described pathways to liberation, freedom and beatitude – Trans.]

40. Art-history has unfortunately not produced a specific study of the significance of the face in the different historical epochs. For such a study would show that these 'degradations of the face' also existed in earlier times. In Mannerism, for example, a period in which not only all faces tend to look the same, but are also depicted as far too small (this is notorious, for example, in Bassano and Greco). Unless the painting is explicitly a portrait, the same degradation of the face is also at work in the Baroque

style, for the transformation of the human into a vividly fluttering being while embellished in elaborate garments is a much more important motif than the face. This is not the place to investigate whether the face was at that time also being devalued because of two de-individualising forces that sought to consume the human. This suspicion seems reasonable, however, because the epoch in question coincided with the burgeoning natural sciences and the Counter Reformation and because in the art of this time a most uncompromising realism was constantly colliding with supernatural exuberance.

41. The sense of time of the person working in conformity with the machine, no longer has to do with the subject's sense of time which (as all modern philosophies of time deem to be self-evident) irreversibly moves forward into the future. The machine-worker sense of time is cyclical. Time is felt as consisting of small-durations, which are co-extensive with the continually re-commencing, identical action of the machine. This is hence time as it is measured by the second hand on a clock, although its full revolutions are in this instance not registered as minutes and hours as they usually are. Only when starting work at the machine do the repetitions 'swim along' in the broad advancing stream of time; and only during this initial phase are they perceived as being boring. That is, they are boring as long as the irreversibly advancing flow of time has not yet lost its hold, which happens after a short while. Once work has become attuned to the machine, the repetitions no longer appear to follow each other in succession. They no longer form a chronological series, as little as one breath of air is registered as following another by the one who breaths (because he or she is always breathing but never knows the number of the breath of air being taken). It is at this point that the repetitions stop being boring. With this I naturally do not want to suggest that they become interesting, only that they are now devoid of time. In fact, these repetitions are so devoid of time that workers are often surprised by the end of the shift: having been robbed of a sense of time they have the feeling of having just started their working day. This is similar to someone who has just woken from a dreamless sleep and is convinced that they only just went to bed. If temporality is not just understood as one trait of subjectivity amongst others, as it has been since Kant, but is rather seen as the decisive attribute (as is apparent in the title of Heidegger's major work), then the fact that time 'ceases' (*eingehen*) while working at a machine can be seen as a symptom for the 'decease' of subjectivity itself.

Part II

UTOPIA INVERTED: GÜNTHER ANDERS IN THE DIGITAL AGE

Things are in the saddle,
And ride mankind.

There are two laws discrete,
Not reconciled, –
Law for man, and law for thing;
The last builds town and fleet,
But it runs wild,
And doth the man unking.[1]

—Ralph Waldo Emerson

[1] Extract from Ralph Waldo Emerson, 'Ode Inscribed to William S. Channing', *Emerson's Complete Works, Volume 9: Poetry* (London: Waverley, 1893), 71–4, 73.

Chapter 1

Better than Human

Promethean Shame and the (Trans)humanist Project

'On Promethean Shame' opens with four diary entries that establish the immediate context from which the text emerged: the American exile that Günther Anders began in 1936 and which lasted until 1949. In light of the thematic trajectory of the remainder of this book, it is highly fitting that the 'setting' of the essay is California, the very place that is at the centre of the innovations in computing and digital technology that are shaping the present in ever more powerful ways. 'On Promethean Shame' bears traces of the experiences Anders gained in California, not only of the Californian way of life and the odd jobs on which he relied to support himself, but also of the specific academic milieu that was forming around a number of more canonical émigré thinkers and literary authors. Anders's essay presents itself as a defence of the thesis that a new form of shame has come into existence, and the narrative unfolds by offering ripostes to a number of purported objections. As readers, we do not know who might have offered these, or if such conversations actually took place in the manner described. It is clear, however, that Anders maintained the intellectual ties that were formed in the late 1920s and early 1930s in Germany. In fact, as Konrad Paul Liessmann elaborates, one of the earliest records of the ideas that would shape 'On Promethean Shame' and the multifaceted engagements with human obsolescence are the transcripts of a presentation at a Los Angeles seminar convened by Theodor Adorno and Max Horkheimer in August 1942.[1]

Anders opened his contribution to the 'Seminar on the Theory of Needs' by reiterating what I have called the 'Promethean definition' of human beings, which he developed in his pre-war reflections on *Weltfremdheit*, the human's estrangedness from the world: 'Artificiality is the nature of human beings. This means that the demand human beings place on the world from the outset surpasses what it supplies.'[2] While discussing this idea – other participants in the seminar included Bertolt Brecht, the Austrian composer Hans Eisler, Friedrich Pollock, Ludwig and Herbert Marcuse – Anders offers the thesis on which his post-war writings hinge: 'The artificiality of human beings increases in the course of history, because humans become the product of their own products. [...] A discrepancy, a widening gulf opens between the human and its products, because human beings can no longer live up to the demands that their own products place upon them.'[3] This increase of the level of artificiality is at issue in 'On Promethean Shame', as the feeling of inferiority, curtailment and obsolescence that can befall us in the view of technological objects is presented as a turning point in the human relationship to technological artifice: the things we turn to in order to settle in ourselves and combat our estrangedness from the world are becoming unsettling themselves. The technological infrastructure has reached a level of development in which we, and the limited resources of the planet, can no longer live up to the demands that the infrastructure generates. The basic dilemma of our age, as Anders puts it in his 'Theses for the Atomic Age', is that '*we are Inverted Utopians*': 'While ordinary Utopians are unable to actually produce what they are able to visualise, we are unable to visualise what we are actually producing.'[4] Anders builds on this platform in his post-war writings to suggest that an ever more fractured and unequal planet is being created, one on which our lives are becoming ever more 'conditional' as they are being shaped by increasingly capable technological objects and the economic processes with which they are intertwined.

The essay 'On Promethean Shame' can be seen as the record of Anders's shift away from the early 'apolitical' trajectory of his thinking, which was focused on the 'artificiality' of human beings as an existential condition, to an emphatically political and pragmatic approach that seeks to think through the concrete implications and results of our retreat to artifice. As such 'On Promethean Shame' – as Anders indicates himself in his 'Theses for the Atomic Age' – is perhaps the key foundation from which his post-war writings emerge, as the essay introduces the idea of a 'Promethean gradient', the ever steeper and unsurmountable slope that separates 'what we can make' (*machen*) from 'what we can imagine' (*vorstellen*), 'what we can do' (*tun*) from 'what we can feel' (*fühlen*), 'what we can know' (*wissen*) from 'what our conscience can register' (*Gewissen*). This gradient, as Anders elaborates in the introduction to *The Obsolescence of Human Beings 1*, is ultimately a manifestation of the growing 'a-synchrony' of the (lived experience of the) body (*Leib*) and the machines (*das Gerät*) to which it is exposed.[5]

The important role that 'On Promethean Shame' plays within Anders's own body of works would in itself have provided sufficient grounds for choosing this particular essay as a point of departure for an engagement with Anders's thought today. A more decisive factor, however, was the sense of timeliness and originality that struck me when first encountering the uncompromising nature of Anders's discussion of shame. By presenting shame as 'a pain of unfreedom' that is at the heart of our desire to overcome, or at least remodel, our own limitations with the help of (technological) artifice, I believe that the essay arrives at a conclusion that has the potential to make a significant contribution to future engagements with this affect.[6] The play of inhibition and agency, hesitation and impulse, attention and oversight that the feeling of shame animates, mediates our relationship to ourselves, our relationship to each other and the way we treat, value and classify non-human entities. If we do not 'feel' the bodily pull of restraint, we are less likely to contemplate our actions. If an object does not attract us, we are unlikely to give it further attention. Anders brings to this problematic an acute awareness of what he calls the 'plasticity' of feeling.[7] Far from being exclusively bodily in origin, the 'pain of unfreedom' is the involuntary collision with bodily limits that are always already being renegotiated (and negated) with the aid of artifice. If shame, inhibition and scruple put us in touch with the body and our acts, technology, as Anders powerfully illustrates, has a tendency to put us out of touch with our own limitations. As such, technology enables us to do things that we could not do without its help, not only because it amplifies our physical and cognitive power and supplements our body with artificial organs, but also because it *amplifies our emotional detachment*.[8] Increasingly capable technological objects do not merely confront us with our own limitations and thus potentially rivet us to the embarrassingly human existence they augment. They also simultaneously put us out of touch – in ever more powerful ways – with the things they do on our behalf.

The radical nature of Anders's engagement with shame has not remained unrecognised in Anglophone scholarship. In her pioneering essay on the relevance of Anders's thought today, '*O Superman!* Or Being towards Transhumanism', Babette Babich notes that 'Anders's most dissonant insight – vying with anything Levinas argues about the face, with anything Heidegger argues about death and thrownness, and everything [...] that Hannah Arendt writes about natality – is that the whole of our problem with modernity, which means starting with Descartes, begins and ends with our awful *shame* at having been born.'[9] The extrinsic circumstances that mark each human life – the fact that one has to come to terms with the lottery of birth, a body, background, familial and national association all of which one did not choose, and the singularity, inequality and constellation of limitations that define our bodily existence – impose inescapable limits on our self-determination, confidence and freedom. As such, Anders suggests, they can become a cause for shame the more they seem to be negotiable. The project of modernity – as Babich's analysis forcefully delineates – addresses these limitations on self-determination not by turning to fellow human beings, but through an investment in technological artifice. Viewed from Anders's perspective, therefore, the enlightenment no longer looks like a project of human emancipation and democratisation that leads to the rise of social institutions that seek to address human inequality. On the contrary, it is a project which, on a more fundamental level, represents an active turn away from, and a faltering trust in, everything human.[10] This trajectory, to cite the subtitle of Sherry Turkle's influential *Alone Together*, means that as modernity unfolds we seem to expect ever 'more from technology and less from each other'.[11]

This refusal of the 'lottery of birth' is central to a further substantial engagement with 'On Promethean Shame' that has recently been published in English: Michael Hauskeller's *Sex and The Posthuman Condition*. Hauskeller adds to Babich's account that the 'pain of unfreedom' is acutely felt in our limited control over our own sexual desires. Adam and Eve cover their genital area, as Hauskeller suggests in opening his analysis with reference to Augustine's *City of God*, because they want to conceal their lack of control. They seek to disguise the fact, as Anders puts it, that as sexed beings they are subject to longings and thus 'belong' to this desire more originally than they belong to themselves.[12] Our sexuality confronts us with our 'unfreedom' in at least two ways: we cannot choose who, what or when we desire, and we cannot influence directly whether this desire is reciprocated or in line with social (and legal) conventions and mores. As Hauskeller's incisive reading shows, the paradoxical nature of the transhumanist project, which strives to achieve 'biological liberation' by using technology to free us 'from the biological and genetic constraints evolution has placed on us and that now represent impediments to us achieving a good life or other valued goals', is fully revealed in the context of love and sex.[13] Rather than abandoning the body and the associated sexed existence, an express aim of the human enhancement project is to 'fix' love and 'augment' the satisfaction of desire by re-engineering it. 'As it stands,' as Hauskeller paraphrases the transhumanist position, 'people fall in love, and they fall out of love. They feel attached to some people, but not to others. […] They cheat on each other, and they are frequently jealous. As a result, relationships break, children are abandoned, and there's generally a lot of suffering, all because we cannot control our desires and the way we feel about other people.'[14] As bio-, digital and robotic technology advances, solutions to these problems seem to offer themselves. This opens the prospect that people can re-engineer their desires and emotions to make a relationship last, begin or end, while also finding technological means – sex-robots and virtual reality environments – to trick the body into producing an intensity of pleasure that can only be partially attained when two clumsily human lovers have sex. The contradictions, of course, should be plain to see and are fully in line with Turkle's diagnosis. In order to 'fix' love, the prospective partners are encouraged to turn away from each other and tend to their own needs and desires with the aid of technology. Love and sex, it seems, are flawed because they are held back by the human condition. This forces us to reach out to others in hope and trust and demands that we suppress our selfishness if we are to build a relationship. Our pleasure is limited as are our relationships. As a consequence, we should not only be permitted to find full sexual stimulation, but also full protection from suffering a broken heart or the process of mourning that the death of a person dear to us exposes us to. We should be spared, in short, from having to be so embarrassingly and painfully human.

I opened this volume by citing Jean-Luc Nancy's diagnosis that '"technology" is a fetish-word that covers over our lack of understanding of finitude and our terror at the precipitate and unbridled character of our "mastery," which no longer knows either end or completion'.[15] Anders's mode of critique does not merely anticipate and confirm this diagnosis; it is, in Hauskeller's words, 'today more relevant than ever'.[16] The overarching aim of this volume is to help further establish Anders's highly pragmatic and overtly critical approach, and to make a case for more translations of and engagements with his work. I hope, therefore, that the encounter with 'On Promethean Shame' has replicated my own sense of discovery and immediate relevance, the 'shock' even, that Ernst Schraube evokes in '"Torturing Things Until They Confess"' when reflecting on his own feelings when reading Anders for the first time.[17] Before I engage with our own technological environment more directly, I will address the suspicion head-on that both Anders's 'antiquated' moralising language[18] and his cutting critique of technological progress are likely to have evoked: Anders is an old-fashioned technophobe; his thought ultimately advocates the sanctity of human beings. It is reactionary and seeks to preserve 'the idea of human dignity', which, as we can read in *The Obsolescence of Human Beings 1*, Anders's father (the eminent psychologist William Stern) had 'ineradicably implanted' in his son.[19]

In fact, if we turn to Ray Brassier's recent defence of the utopian, transhumanist quest of combatting human limitation, we might be moved to think that Anders is one of this project's most ardent critics. For Brassier, 'Prometheanism is simply the claim that there is no reason to assume a predetermined limit to what we can achieve or to the ways in which we can transform ourselves and our world. But of course, this is precisely what theological propriety and empiricist good sense jointly denounce as dangerous hubris.'[20] Brassier turns to a critique of Jean-Pierre Dupuy's 'Some Pitfalls in the Philosophical Foundations of Nanoethics' which dissects and critically responds to Ray Kurzweil and Terry Grossman's transhumanist rallying cry that 'disease and death at any age [are] a calamity', that they are '*problems to be overcome*'.[21] Dupuy, who cites 'On Promethean Shame' in the course of his discussion, draws on Hannah Arendt and Ivan Illich to outline that finitude is a defining trait of human existence, and he asks if we 'do not lose something infinitely precious' when we perceive finitude as 'an alienation and not a source of meaning'.[22] In Brassier's account of Dupuy's position, this sentiment is represented as a call for imposing limits on the Promethean project of re-engineering the given ('natural') order, a call that is presented as being maliciously motivated by a religious conviction that pain, suffering and death are 'the prerequisites for a meaningful existence'.[23] It is here, I believe, that Prometheanism as conceived by transhumanist positions runs into the fundamental contradiction that is also exposed by Babich and Hauskeller. For, as Dupuy shows, the dream of surpassing limits, the turn away from death, which, as we have seen, Prometheus already signifies in Greek tragedy – 'I stopped mortals from brooding on death', 'I made to lodge within them blind hopes' – is in itself the inextricable mark of an existence exposed to finitude.[24]

The question, therefore, is not whether one is *for* or *against* finitude and most certainly not whether one is *for* or *against* pain – it is perhaps needless to say that Dupuy does not advocate suffering as an end in itself as Brassier seems to imply. The question, rather, is how we can find ways to live with our own finitude and limitations as these boundaries become ever more malleable and indefinite. This is not an individual or an exclusively scientific task, but a collective one to be met in all spheres of human activity, for finitude is the basic condition we all share. This position, of course, is easily presented as being conservative, because it focuses on the pain, suffering and the violence that *is unleashed* in combatting finitude, rather than being transfixed by the utopian endpoint in which these human 'problems', to use Kurzweil's term, no longer exist. For Dupuy, therefore, the dream of a beyond finitude is not only the mark of the shame of being born that Anders introduces us to, it is a desire that can only emerge in response to the experience of limitation. The challenge, and Dupuy here echoes Nancy, is to engage in a renewed thinking of finitude, not only to respond to the technological advances that promise to have the potential to re-engineer (certain) limits indefinitely for the select few who can afford to re-engineer themselves, but also to account for the motivations to overcome these limitations in the first place. For, even if we disregard the fact that the endpoint of transhumanist scientism – an existence devoid of pain, suffering, death – might not be as achievable as Kurzweil, Brassier and 'Accelerationist' positions imagine (even if science were to be unshackled from the restraining chains of capitalist competition and patent law),[25] Brassier seems entirely unwilling to probe his own source of hope. While accusing Dupuy of holding on to a religious paradigm that sanctifies suffering, Brassier, in line with the transhumanist movement, presents us with *a life beyond limitation and death*, and nothing could be more 'religious' than this. Such scientism and unbridled faith in progress are hence – and this is Anders's basic position – not as motivated by reason as one might like to think, and as such, they are not immune to the fanatical excesses (and forms of violence) that we like to ascribe to religious fundamentalism. This form of violence, of course, does not openly play itself out in the resurgence of archaic forms of violence that are a trait of 'religious' violence today. As I show in the remainder of this book, the violence that accompanies human enhancement and quantification comes to expression, rather, in the discrepancies that Anders addresses when discussing the concentration of power and the sanitisation of its workings that takes effect as the Promethean gap between our bodily perception and the objects we create widens. This violence, then, is anaesthetic rather than aesthetic. As such, it does not draw attention to itself, but is self-concealing and decidedly un-sensational. If anything, then, Dupuy is not a 'critic' of Prometheanism, but a thinker who addresses the 'Promethean condition' of existence, a condition that it becomes ever more urgent to think as our technological objects give us godlike powers. I believe that Anders's thought can make a major contribution towards the further development and pragmatic reorientation of approaches that do not 'cover over our lack of understanding of finitude'.

The object-oriented nature of Anders's engagements opens us to modes of critique that remain focused on the actual realities we are establishing, rather than turning our attention onto future worlds and an endpoint – life without limitation – that might not ever be attained. If we turn to Section 7 of 'On Promethean Shame', where the phenomenon of 'hubristic self-degradation' is introduced, it becomes clear that Anders is neither for nor against technological progress, but intent on thinking through the consequences of Prometheanism. In line with Brassier's definition, Anders suggests that there 'is no reason to assume a predetermined limit to what we can achieve or to the ways in which we can transform ourselves and our world'. For Anders, however, this means that 'what exists, does not have to be'; the only measure we have to evaluate our 'progress' is '*humanity itself*'. Put otherwise, the life, social and environmental conditions we are creating while pursuing progress should be our principal concern, because unlike the target condition, these provide the concrete realities with which we have to deal. The very advent of 'Promethean shame' is taken as the sign that the world we have created already definitively surpasses our limited comprehension. In the course of history we have collectively established an inverted utopia, a world built by humans that nevertheless has left us unable to visualise the effects of the products we make. The very premise that motivates 'On Promethean Shame' is that there are no predetermined limits to what human artifice can achieve. There is, however, an inflection point when the pursuit of good human intentions and technological progress in the pursuit of happiness, health and self-determination create living conditions that are no longer tolerable. 'It may be true', as Anders puts it with regards to human engineering, 'that humans are doing all this in the blind hope, or at least on the pretext, that this self-reification will reduce or even abolish all other forms of human frailty and degradation. But what counts are not reasons, but, rather, what is actually done: in order to combat the limitation of their liberty, humans limit themselves in a new way.'[26]

What we are in danger of establishing while combatting our limitations, Anders suggests, is a 'world without us'. This is not merely a world that is devoid of life as an endpoint of total nuclear war or climate change, but a world in which human contributions have been devalued to such an extent, and our collective ways of life are controlled by the small minority that decides on (and profits from) the future course of technological development, that 'we' the figurative '99%' who remain excluded from these processes no longer find the space and opportunity to settle in a world increasingly belonging to and shaped by others.[27] Precisely because artifice is the essence of man, as Anders can help us contemplate, the one thing we must learn now is the decidedly non-essential, fragile 'art' of being human. Anders, as I see it, does not oppose technological progress, but the regression that inevitably happens to the sphere of human interaction and moral cognition as we turn to machines (rather than fellow humans) while combatting our limitations and attempting to solve our problems. Today, as Anders puts it at the outset of the second volume of *The Obsolescence of Human Beings*, 'It is not enough to change the world. Humans do this anyway. [...] We also need to interpret this change, in order to change it. So the world no longer changes without us, and does not end up being a world without us.'[28] As part of this task, Anders repeatedly calls for a 'Critique of Pure Feeling', not one aiming at 'reaching a moral verdict', but a critique in the Kantian sense – a critique, therefore, that seeks to trace the limited performance and plasticity of our feelings in a world populated by ever smarter and more autonomous machines.[29] It is to this problematic that I turn to now.

NOTES

1. See Konrad Paul Liessmann, 'Die prometheische Scham – der Mensch und seine Geräte', *Günther Anders zur Einführung* (Hamburg: Junius, 1988), 33–50, 33–4. This seminar and its relation to Anders's later thought and the Frankfurt School of Critical Theory is discussed at length in German language scholarship. Besides Liessmann, see, for instance, Werner Fuld, 'Zwischen Film und Bombe. Die Kontinuität des Andersschen Denkens', *Günther Anders Kontrovers*, ed. Liessmann (München: C.H. Beck, 1992) 114–23, 121f. The transcripts in question have been compiled and published under the title 'Diskussionen aus einem Seminar über die Theorie der Bedürfnisse', *Max Horkheimer: Gesammelte Schriften Band 12: Nachgelassene Schriften 1931–1949* (Frankfurt am Main: Fischer, 1985), 559–86.

2. Anders explains the imbalance of supply and demand as follows in a 1949 lecture held at the New School in New York: 'Man, however, and *here* we *come* to one of *the vital* conditions *of culture*, by principle *asks for more than nature can supply*. And since this "*more*" cannot be supplied by the blue sky, he is making this product himself. [...] This fact of *un-balanced* demand and supply-proportion means [that man] *not only needs or demands* more than what he is *seeing*; it means, that he is *seeing more* than what he *demands*. He demands tables and chairs – yet he sees – as supply offered by Nature, trees; he demands steak, yet what he finds are living beings; he demands bread, yet what he sees or finds, is soil and grain. Thus he sees [the] world, *as not satisfying his demand*, he sees it as *itself*; he sees *nature as nature, because he needs or wishes more than nature offers*. Insofar as we can say "nature," *we are more than nature*' (English in the original). See Anders, 'Kulturphilosophie' (unpublished manuscript) as cited in Christian Dries, *Die Welt als Vernichtungslager: Eine kritische Theorie der Moderne im Anschluss an Günther Anders, Hannah Arendt und Hans Jonas* (Bielefeld: Transcript, 2012), 47.

3. See 'Diskussionen aus einem Seminar über die Theorie der Bedürfnisse', 579.

4. Günther Anders, 'Theses for the Atomic Age', trans. Günther Anders, *Massachusetts Review* 3.3 (Spring 1962), 493–505. I am citing from the reprint of the essay that is part of *The Life and Work of Günther Anders*, 187–94, 189.

5. Günther Anders, *Die Antiquiertheit des Menschen 1:* Über *die Seele im Zeitalter der zweiten industriellen Revolution* (München: C.H. Beck, 2010), 16.

6. Anders, 'On Promethean Shame', (ch1) 93*.

7. Anders, *Antiquiertheit 1*, 271ff.

8. Anders frequently illustrates this emotional detachment along the following lines when introducing his notion of a 'Promethean Gradient' in the context of nuclear weapons and air war: 'Our abilities and faculties each have their own relationship to magnitude, measurement and scale. Example: *We can easily plan and execute the destruction of a major city with the aid of instruments of annihilation we have produced ourselves. But we can only inadequately imagine the effect. And yet the indistinct image we can picture:* smoke, blood and ruins is still quite a lot in comparison to the tiny quantum of emotion and responsibility we can muster in view of this thought. Each faculty hence has its own limit beyond which it can no longer operate. We can kill thousands, but at best we can imagine ten dead people, but we can cry over and mourn at most one person (at a time). [...] A city full of dead people remains a mere word to us.' See Anders, *Antiquiertheit 1*, 267.

9. See Babette Babich, '*O, Superman!* Or Being Toward Transhumanism: Martin Heidegger, Günther Anders, and Media Aesthetics', *Divinatio* 36 (2013), 42–99, 75–6.

10. Section 3 of 'On Promethean Shame' evokes a passage from Nietzsche's *Zarathustra* that addresses humanity's evolutionary past as a shameful heritage: 'You have made your way from worm to human, and much in you is still worm.' Zarathustra's famous conclusion, that the 'human being is something that must be overcome' seems to provide the basic motif that Anders's take on the dream of modernity develops, for Zarathustra continues by introducing the shame of having to live as a human being: 'What is the ape to a human? A laughing stock or a painful embarrassment. And that

is precisely what the human shall be to the overman: a laughing stock or a painful embarrassment.' See Friedrich Nietzsche, *Thus Spoke Zarathustra*, trans. Adrian del Caro (Cambridge: Cambridge University Press, 2006), 5–6.

11. See Sherry Turkle, *Alone Together: Why We Expect More from Technology and Less from Each Other* (New York: Basic, 2011). The circumstance that the turn to technology often concretely means a turn away from humans is a central motif to further influential critiques of our current technological environment. See, for instance, Jaron Lanier's *You are not a Gadget* and Evgeny Morozov's *To Save Everything Click Here.*

12. See Michael Hauskeller, *Sex and the Posthuman Condition* (Basingstoke: Palgrave Macmillan, 2014), 47–8; Anders, 'On Promethean Shame', (ch1) 67*.

13. Julian Savulescu and Anders Sandberg, 'Neuroenhancement of Love and Marriage: The Chemicals between Us', *Neuroethics* 1, 31–44, 41 (as cited in Hauskeller, *Posthuman Condition*, 49).

14. Hauskeller, *Posthuman Condition*, 48.

15. Jean-Luc Nancy, 'A Finite Thinking', *A Finite Thinking*, ed. Simon Sparks, trans. Edward Bullard, Jonathan Derbyshire and Simon Sparks (Stanford: Stanford University Press, 2003), 3–30, 25.

16. Hauskeller, *Posthuman Condition*, 42.

17. Ernst Schraube, '"Torturing Things Until They Confess": Günther Anders's Critique of Technology', *Science as Culture* 14.1 (2005), 77–85, 77.

18. It is hard not to agree with Liessmann's point that Anders's reflections on human obsolescence can seem decidedly antiquated because their visionary approach is carried forward in a language marked by the vocabulary of early twentieth-century German thought, cultural pessimism, and a 'prophetic-apocalyptical tone that can be unbearable, especially for young people'. See Liessmann, *Günther Anders: Philosophieren im Zeitalter der technologischen Revolution* (München: C.H. Beck, 2002), 184.

19. Anders, *Antiquiertheit 1*, v.

20. Ray Brassier, 'Prometheanism and its Critics', *The Accelerationist Reader*, ed. Robin Mackay and Armen Avanessian (Berlin: Merve, 2014), 467–87, 471–2.

21. As cited in, Jean-Pierre Dupuy, "Some Pitfalls in the Philosophical Foundations of Nanoethics," *Journal of Medicine and Philosophy* 32 (2007), 237–61, 248. Kurzweil's ideas are also a central point of criticism in Babich and Hauskeller.

22. Ibid., 249. See Dupuy's *The Mark of the Sacred* for a book-length discussion of this question.

23. Brassier, 'Prometheanism', 481.

24. Aeschylus, *Prometheus Bound*, trans. A. J. Podlecki (Oxford: Aris & Phillips, 2005), 93.

25. A key motivation of the 'accelerationist' movement, which, as the name indicates, calls for a politics of speeding up technological progress to address the imbalances, damage and inequalities capitalist modes of production inevitably produce, is the liberation of techno-scientific labour from capitalist models of intellectual ownership. See Alex Williams and Nick Srnicek, '#Accelerate: Manifesto for an Accelerationist Politics', *The Accelerationist Reader* (Berlin: Merve, 2014), 347–62, 355.

26. Anders, 'On Promethean Shame', (ch1) 50*.

27. See Anders, 'On Promethean Shame', (ch1) 33*. Jaron Lanier, in line with some key premises of Anders's thought presents this problematic as follows while introducing *You Are Not a Gadget*: 'We [Lanier is here speaking in the name of "Technologists"] make up extension to your being, like remote eyes and ears (webcams and mobile phones) and expanded memory (the world of details you can search for online). These become the structures by which you connect to the world and other people. These structures in turn can change how you conceive of yourself and the world. We tinker with your philosophy by direct manipulation of your cognitive experience, not indirectly through argument. It takes only a tiny group of engineers to create technology that can shape the entire future of human experience with incredible speed. Therefore, crucial arguments about the human relationship with technology should take place between developers and users before such direct manipulations are designed.' See Lanier, *You are Not a Gadget* (London: Penguin, 2007), 5–6.

28. Anders, *Antiquiertheit 2*, 5.

29. Anders, *Antiquiertheit 1*, 269.

Chapter 2

The Punishment of Prometheus

On the evening of 6 August 1945, President Harry Truman issued a public statement from his White House office: 'Sixteen hours ago an American airplane dropped one bomb on Hiroshima and destroyed its usefulness to the enemy. That bomb had more power than 20,000 tonnes of TNT. It had more than two thousand times the blast power of the British "Grand Slam" which is the largest bomb ever yet used in the history of warfare.' It is estimated that the explosion this bomb generated instantaneously killed over 110,000 people (many tens of thousands more than there are words in this book). In the course of the statement Truman became more specific about what had been unleashed against the city that morning: 'It is an atomic bomb. It is a harnessing of the basic power of the universe. The force from which the sun draws its power has been loosed against those who brought war to the Far East.' He added a warning that 'the end is not yet', that further and 'more powerful forms' of the explosive device were already being readied to be deployed.[1] The day on which this Promethean fire was for the first time directed against human beings is a central co-ordinate of Anders's philosophy of technology.[2] With the advent of nuclear weapons, as Anders explains, 'the Promethean dream of omnipotence' at long last came true. This omnipotence, however, arrived in an inverse form, not as the power to 'create out of nothing', but as the ability to categorically destroy and 'reduce to nothingness'.[3]

Although the mastery of the atom left us 'anatomically unchanged', Anders declared that humanity turned into a 'new species' that has fulfilled its Promethean destiny by attaining nuclear weapons: 'We are Titans, no longer what until today humans have called "human."' This titanic posthumanity is evoked, however, only to impress on the reader what this triumph signifies: 'Because we are the first human beings with the power to unleash a world cataclysm, we are also the first to live permanently under its threat.

Because we are the first Titans, we are also the first dwarfs, or whatever we may call beings such as ourselves who are mortal not only as individuals, but also as a group, and who are granted survival only until further orders.' These apocalyptic lines, with which 'Reflections on the H Bomb' opens, powerfully illustrate the tensions that 'On Promethean Shame' introduces on a personal level by transposing these onto a global scale: *the more our technological power grows, the smaller we become; the more unconditional and unlimited the capability of machines, the more conditional our existence; the more machines connect us by virtue of their very existence, the more we are also singled out as being expendable and inadequate.*

Almost exactly seventy years after Hiroshima, and over twenty-five years after the non-apocalyptic conclusion of the Cold War, Anders's words seem to belong into a different time. This makes it all too easy to miss the inference that gives Anders's thought its powerfully critical edge. For Anders, 6 August 1945 represented an absolute and final epochal shift. On that day, Hiroshima became a 'world condition' (*Weltzustand*) and a new age began. 'However long this age may last, even if it should last forever, it is "The Last Age": for there is no possibility that its "defining aspect," the possibility of our self-extinction, can ever end – but by the end itself.'[4] What Anders is highlighting here, is not merely the irreversibility and inevitability of technological progress, but the radically changed nature of human existence in view of increasingly complex technological objects. It has become fashionable to herald the democratising potential that new information technologies can bring. In *The New Digital Age*, for example, a book imbued with the optimism of the Arab Spring, Google's executive chairman Erik Schmidt and his colleague Jared Cohen argue that inexpensive media devices have the potential to lead us towards 'digital empowerment', meaning that 'authoritarian governments will find their newly connected populations more difficult to control, repress and influence, while democratic states will be forced to include many more voices (individuals, organisations and companies) in their affairs'.[5] In a stark and perverse contrast to this, Anders's reflections on our existence in the shadow of the bomb highlight that the dream that technological progress could bring humanity closer together began to be perfectly realised with the build-up of nuclear arsenals after the Second World War. For despite the highly secretive and exclusive nature of the research and decision-making process leading to its existence, the bomb created the first all-inclusive, absolutely unconditional 'we'. Regardless who made the bomb and how or why it was made, once it came into existence everybody had to begin a life in the shadow of this particular technological know-how. Despite, or rather precisely because of, 'our' newly attained power, every single individual simultaneously became impotent and disempowered in a new way. The bomb created the first World Wide Web as its creation connected us all.[6] With the power to exterminate us all to the same absolute degree, nuclear arsenals act as the ultimate levellers of humankind, even though they simultaneously concentrate an unprecedented amount of power onto single agencies. 'Protagoras saying "man is the measure of all things" has lost all its validity,' Anders concludes in *Heresies*, '"for our measures" have become the machines and products that we have made. What is measured, is merely if we are still capable of living up to the measure they set, for this decides if we may still "be" or if we are allowed to "be no more."'[7]

For Anders, the inherent potential nuclear weapons have to create a world on which no human life remains, illuminates how complex technological objects also liquidate humans by making us superfluous and turning off our critical and emotional faculties. Complex technological objects, as Anders puts it, harbour 'worlds without us' the very moment they also enhance our agency, ability and power. As Jean-Luc Nancy recently remarked, Anders's 'stark but little heeded lesson, too little mentioned in any case in the world of philosophers', is condensed into the title of one of Anders's books: in a world populated by complex machines, '*Hiroshima is Everywhere*'.[8] In the remainder of this book, I reiterate this provocation in the context of the digital age. In Anders's writings, therefore, the atom bomb is *not an exceptional machine*, but assumes the position of being *the exemplary* technological object. As the most powerful and deadly machine, and a product of highly compartmentalised and specialised labour processes, the atom bomb gives us a glimpse of the worlds that are configuring themselves around increasingly capable, autonomous and connected machines. And I use the plural 'worlds' here because each new technological object also conceals within a world it can actualise in its image, and not all of these worlds, as the inherent potential of the bomb reveals, can be human worlds.[9] There is no shortage of utopian and dystopian visions that imagine future worlds, but before I turn to the potential of Anders's thought today in more detail, it is worth contemplating this impulse to turn towards imagined futures, as this already informs the earliest critical engagement with technology and the realm of artifice in the Western imaginary: Prometheus.

Already in its oldest surviving version dated to the eighth century BC, the story of Prometheus raises the spectre of future disaster: 'You are pleased at having stolen fire and outwitted me,' Hesiod's Zeus tells Prometheus, 'a great calamity both for yourself and for humans to come.'[10] Amid media speculation about artificial intelligence, daily reminders of global warming, the machine assisted rise of globalised financial and corporate capital, and the vague awareness that nuclear arsenals have outlived the Cold War, Zeus's warning about the calamitous force of technology brings to mind a number of (reassuringly intangible) referents. The very antiquity of the warning serves to illustrate, moreover, that the technological apocalypse

Zeus gestures towards seems to have been deferred for millennia. Nothing is more antiquated, it seems, than predicting a future end brought about by new artificial skills.

It is against the backdrop of this impulse to look ahead into imagined futures when responding to the way in which digital technologies are reconfiguring our lives that I have proposed taking a *progressive step back* to Anders's philosophy of technology. In the next chapter, I begin my discussion of digital technology by taking a step back to the late 1920s, as the introduction of analog wire technologies is arguably the germinal moment of the emotional detachment that Anders invites us to think. This backward turn also servers to accentuate that viewed through Anders's "Promethean perspective" of technology, human obsolescence – and the calamities Zeus announces – haunt us from the distant past, and not from the future in the form of 'smart' devices that have the capacity to replace us. As 'indeterminate' beings who fundamentally rely on the artificial skills and technologically mediated abilities we acquire after birth, *we are born obsolete*. To be human means to live with a compulsion to leave behind the Epimethean condition of being 'naked, unshod, without bedding or weapons' that marks our arrival in the world. Once born, we are inevitably propelled into the Promethean realm of artificial skill and technological supplementation. Technology, as I showed in the introduction, does thus not merely name the constellation of 'things and gadgets' we surround ourselves with, but *a space of possibility* that shapes what we can become by configuring the way we act, think, feel, speak and encounter ourselves and other entities. Anders's thought unfolds between the two extreme poles Prometheus and Epimetheus signify. *Prometheus names the realm of doing* (the 'art' of making, experimenting, planning ahead – what the Greeks called *techne*) and *Epimetheus the realm of bodily restraint, unfreedom and limitation* (the art of hindsight, inhibition, feeling and conscience – arts that Plato introduces as *dike* and *aidos*). The underlying thesis of 'On Promethean Shame' is that 'Prometheus has won too triumphantly', that we as Promethean beings are beginning to feel inferior, humiliated and degraded by the increasingly capable technological objects that populate our lives. It is worth establishing, therefore, which conflict Prometheus is winning so triumphantly. As a means of transition, I now briefly return to the myth of Prometheus, and the disharmony between the Titan personifying 'foresight' and Epimetheus, his idiot brother who stands for 'hindsight' and who thus always *cares too late*.

The warning with which Hesiod has Zeus announce two calamities, one awaiting Prometheus in the form of his punishment for tricking the gods, and the other awaiting humankind as a consequence of accepting his gift, is shaped by this dynamic of foresight and hindsight. Both calamities, I suggest, have to do with our bodily existence and the manner in which this bodily reality can make us collide with ourselves when we feel affected by pain, hunger, sexual desire, shame, and other bodily motions that force our thoughts and attention back onto ourselves. In fact, the story as such can be read as a reflection on the nature of our bodily experience. It is Epimetheus, who gives a particular experience of the body: *exposure*. His error leaves us 'naked' thus exposing us to the elements, but also to our limitations, restrictions and needs. Prometheus, in contrast, is the Titan who removes us from bodily limitation. As the personification of 'foresight', 'trickery' and 'planning ahead', Prometheus signifies disembodiment and bodily absence. His is the abstract realm of thought; his gift of artifice couples our ability to distance ourselves from our bodily reality in imagination with an ability to re-modulate our bodies by the means of technological intervention.

The basic example of clothing can serve to illustrate, however, that Promethean artifice can mobilise the trickery of machines to combat bodily exposure only to trick us into a new experience and perception of the body. As Promethean beings, we 'forget the naked skin' underneath our clothes, as Max Scheler explains this in a phenomenological account of shame.[11] In fact, this naturalisation of the cover is the prerequisite for our experience of nudity: being clothed needs to be intuitively apprehended as the natural and normal condition, before nudity can assume any particular significance and revelatory force. Rather than simply removing us from the body, Promethean artistry hence alters the way we perceive, experience and relate to our bodily existence. Yet regardless of how it is modified, bodily feeling will play its part in policing the boundary of what is natural and normal. For, once clothing is 'naturalised' it seems fully counter-intuitive to regard it as artificial, or even as a 'machine' that 'tricks' nature – to reiterate Hadot's definition[12] – out of our physical exposure to the elements only to trick us into a new bodily experience: nudity. This means, as Anders reflects in a diary entry from 10 May 1949, that 'the story recounted in *Genesis* presents everything the wrong way round. Adam and Eve could only discover (*entdeckt*) their nudity, because they had already covered themselves up (*bedeckt*) prior to this.' Anders provocatively continues the inversion of *Genesis* by turning the fall into an ascendency: 'It is only now, with the discovery of nudity, that the gates to Paradise open themselves.'[13] The general example of clothing further illustrates one of the basic ideas of Anders's thought, namely that technological additions remodel our sense of normality, that they make the '*a posteriori*' (that which is added to the body after birth) appear as if it were an '*a priori*' (there before, necessary and concealed within), a process through which the 'artificial world' presents itself as if it were 'nature'.[14] Put otherwise, artifice only seems artificial until it has been successfully incorporated into one's bodily sense of self, only until we are used to it to such an extent that it has slipped from perception. The conflict, then, in which Prometheus has entangled us, is one between bodily absence and bodily presence, and it is this tension that sets the stage for the two calamities that Zeus predicts.

The first calamity Hesiod predicts affects Prometheus himself. It announces his punishment. Zeus has Hephaistos forge unbreakable chains to bind Prometheus to a freezing rock at the edge of the known world. Condemned to painful solitude but yet displayed in shame as an example to others, he is nakedly exposed to the elements without protective cover and is promised countless years of torment. The chains forged by Hephaistos do not remove Prometheus from his body but rivet him to it, without a means of escaping its 'presence'. Every morning as the sun rises to begin to burn his skin, an eagle sinks its talons into Prometheus's flesh to peck at his liver. Every sundown the lacerated organ regrows while the Titan is immersed in a night so cold that he can never find enough comfort to sleep, but is instead racked by worry about the torments the next day will bring. Amid this gruelling cycle, the quality of foresight that Prometheus embodies adds to his torment: 'I have clear and thorough knowledge of all that is to come; no unexpected misery will come to me.'[15] The corporeal nature of the punishment continuously reiterates, however, that there exists a qualitative difference between pain being anticipated and pain being felt. What Zeus displays as Proemtheus's shame, therefore, is the limited and disembodied nature of the Titan's sight. Although Prometheus knew that he would be punished, this knowledge did not inhibit his deed for he could not understand the meaning of pain. Only with hindsight, only when the Eagle's beak splits his skin to injure his most vulnerable part, does this knowledge begin to sink in. The conclusion we might draw from this is a central motif of Anders's philosophy of technology: disembodied knowledge is a form of ignorance, no knowledge at all.[16]

This aspect of the myth is isolated by Arthur Schopenhauer, who turns Prometheus's punishment into a parable for human experience itself:

> Prometheus personifies the particularity of human foresight and the care for that which lies ahead (*Vorsorge*). […] As a price for this privilege, however, humans have to suffer the torment of *worry and concern*. This torment is the vulture that emaciates the liver of bound Prometheus. Epimetheus, who was probably invented retrospectively as a logical addition to the myth, represents this quality of *hindsight* and *caring after the fact* (*Nachsorge*), for this is the payment we have to exert for improvidence and absentmindedness.[17]

The punishment Zeus inflicts hence encapsulates the discrepancy between foresight and hindsight, which is also a discrepancy between disembodiment and embodiment. The former is marked by an absentmindedness, for it lacks the understanding that only lived experience can bring. As creatures made by Prometheus, we do not only inherit his foresight, the desire and ability to shape the course of the future by mobilising his gifts, we are also marked by his lack of understanding and blindness. We animate the trickery of machines to experiment freely, unbridled by care. By trial and error we confidently aim at anticipated and hoped for results. We can do so, however, because it is only hindsight that will reveal the full extent of the consequences, and these consequences might have made us hesitate or reconsider our action, if only they could have been properly understood. Put otherwise and through the lens of Anders's early thought, as Promethean beings we learn by experience, but this very fact also accentuates a rift – *Weltfremdheit* – for experience will permanently exhibit the lack of understanding that hindsight reveals: 'We remember therefore we are,' as Anders puts it. Memory and experience, however, cannot definitively eradicate the originary rift that opens us to consciousness of self, for we also remember that we did not know what we would become.[18]

The parable of Prometheus's punishment suggests that the imbalance between disembodiment and embodiment, and foresight and hindsight, that is, the imbalance between carefreeness and anguished concern, and abstraction and concretion (i.e., experience) marks us essentially. Without this imbalance we would have no experience and would have never done anything – our thoughts would know no projects, no unlived moments and would inhabit no tomorrow, our hands would be tool-less and would therefore not be hands, for they would never be directed at the accomplishment of a task. The planet would not know the trickery of 'human' artifice that turned it into a world shaped and claimed by human beings. Without the mistake of Epimetheus and Prometheus's redemptive gifts, we would thus live like the idealised animals that the myth imagines – perfectly equipped and adapted to our natural environment, with no need or desire to build ourselves a second, artificial world to inhabit. In a time, however, in which we are committed to 'the largest experiment involving anarchy in history' – Eric Schmidt and his colleague Jared Cohen are referring to 'the Internet' – Promethean carelessness is taking on an entirely new dynamic. And this leads on to the second calamity Hesiod has Zeus predict, the one reserved for humankind.[19]

Tempting as it is to read this further calamity as a technologically induced apocalypse, piecing together the timeline of the myth across its various versions reveals that the prediction does not speak about the end of the human world but its beginning. It is Plato who introduces us to the events that immediately follow after Prometheus bestowed technology onto us:

> [Man] soon used his skill to develop articulate speech and words for things, and found himself housing, clothing, footwear, bedding, and food from the earth. [...] [Humans] then sought to group themselves together, and protect themselves by founding cities. And when they were grouped together, they started treating each other unjustly – lacking the art of politics as they did – and consequently split up again and went on being destroyed [by wild beasts].[20]

After passing on his gifts, Prometheus hence has to witness how the human adventure almost comes to an end before it begins. These proto-humans 'started to treat each other unjustly' because they as yet lacked *aidos* (humility, respect and shame) and *dike* (justice and the rule of law).[21] The technological is here marked by an originary lack of inhibition. The sphere of artificial skill (which includes the use of language) does not only lay the foundation for human sociality, it also sets in motion desires, calculations and visions of the world that are inherently destructive and antisocial. In the account given in Plato's *Protagoras*, it is only Zeus's last minute intervention that saves humankind from destruction by supplying the *humanising arts* of respect and justice, arts which give us the ability to live together and care for other entities. These are the arts of *non-use*, *not doing* and *not saying* what artifice gives us the power to do – insult, injure and destroy each other, but also the ability to seemingly rely on ourselves, thus closing ourselves off from encounters with others in ways that can leave us alone, isolated and exposed to the hostilities of a predatory, neoliberal world. Technology, as Sherry Turkle so powerfully puts it, increasingly invites us to live 'alone together', connected by devices that have the capacity to hold us apart with ever more fateful consequences. Read through Plato's version of the myth, the 'calamity' Hesiod evokes is not apocalyptic in the sense that technology may destroy the earth or humankind. What is calamitous, rather, is that Zeus has to intervene to furnish us with these arts in the first place, that we lack the qualities required to not do each other harm. Anders's writings, imbued with the 'fateful tone of the tragic', confront us with the thesis that the complexification of technological objects is also enhancing the antisocial potential of technology in ever more powerful ways.[22] This emotive detachment does not merely turn us into internet trolls; it rather points to a situation in which the cultivation of humanising arts is being all but abandoned as we struggle to keep up with a world populated by ever more capable technological objects. To begin to illustrate what is at stake in such reflection, the final part of this book discusses the disembodying, 'anaesthetic force and violence' of (digital) technology by turning to what Anders calls the 'natural' environment in which the relationship between humans and machines plays itself out: the world of work.[23]

NOTES

1. Harry Truman, 'Statement by the President of the United States (6th of August 1945)', accessed via http://www.trumanlibrary.org/whistlestop/study_collections/bomb/large/documents/index.php.
2. Anders describes the significance of the 6th of August 1945 to his own work in his 1982 Preface to *Hiroshima is Everywhere*: 'For many years I could not emerge

from the stupor that took hold of me when listening to the famous radio-broadcast on the 6th of August 1945. [...] Only in the early 1950s, in 1952 or 1953, long after returning from my American exile, I managed to take the first unsteady steps. While sitting under a nut tree in a pleasant garden close to Leipzig, I forced myself – I had practically issued myself with an ultimatum – to write down a few sentences on this "subject." I had persuaded myself that this subject was the most important of all, and that in comparison to "engaging" with this "topic" turning to any other would be inexcusable, childish even.' See *Hiroschima ist überall* (München: C.H. Beck, 1982), xi.

3. Anders, 'Reflections on the H-Bomb', trans. Norbert Guterman, *Dissent* 3.2 (Spring 1956), 146.

4. Anders, 'Theses for the Atomic Age', 187.

5. See Eric Schmidt and Jared Cohen, *The New Digital Age* (London: John Murray), 7. This position is expanded on in the course of this evangelical text, see esp. 'The Future of Identity, Citizenship and Reporting', 33–81.

6. It is an interesting aside that the internet owes its current form to cold war fantasies. As Lanier explains: 'The internet was [...] originally conceived during the Cold War to be capable of surviving nuclear attack. Parts of it can be destroyed without destroying the whole, but that also means that parts can be known without knowing the whole.' See *You Are not a Gadget*, 123.

7. Anders, *Ketzereien* (München: C.H. Beck, 1996), 119.

8. See Nancy, *After Fukushima*, 19.

9. This multiplication of worlds is the subject of a discussion between Jean-Luc Nancy and Aurélien Barrau which has shaped this project. See esp. Jean-Luc Nancy, 'Of Struction', in *What's Theses Worlds Coming To?* trans. Travis Holloway and Flor Méchain (New York: Fordham University Press, 2015), 42–58.

10. Hesiod, *Theogony and Works and Days*, trans. M. L. West (Oxford: Oxford University Press, 1999), 38.

11. Max Scheler, 'Über Scham und Schamgefühl', in *Schriften aus dem Nachlass Band 1: Zur Ethik und Erkenntnislehre*, ed. Manfred S. Frings (Bern: Francke, 1957), 65–154, 75.

12. See intro, 2* above.

13. Anders, *Tagesnotizen* (Frankfurt am Main: Suhrkamp, 2006), 65.

14. Anders, *Antiquiertheit 1*, 311.

15. Aeschylus, *Prometheus Bound*, 83.

16. This is discussed, for instance, in a reflection on why we accept the existence of nuclear weapons despite knowing their potential: 'That we "know" what consequences a nuclear war would have cannot be denied. But we only "know" it. And this "only" indicates that our "knowledge" is in the closest proximity to "not-knowing," that is, to ignorance.' Anders, *Antiquiertheit 1*, 269–70. On this point see also my 'Desert Ethics: Technology and the Question of Evil in Günther Anders and Jacques Derrida', *Parallax* 21.1 (2015), 42–57, 47f.

17. Arthur Schopenhauer, *Parerga und Paralipomena II: Zweiter Teilband* (Zürich: Diogones, 1977), 452–3.

18. On this point see Anders, 'The Pathology of Freedom', esp. 297–302.

19. Schmidt and Cohen, *The New Digital Age*, 3.
20. Plato, 'Protagoras', 158.
21. Plato, 'Protagoras', 158–9.
22. See Werner Reiman, 'Nihilismus und Scham', *Günther Anders Kontrovers*, ed. Liessmann, 57–71, 62.
23. Anders, 'On Promethean Shame', (ch1) 81*.

Part III

ANAESTHETIC LIVES: JOYFUL SURRENDER TO PAINFUL OBSOLESCENCE

In 1928, Siegfried Kracauer produced a pioneering reportage about the effects of rapid technological change. *The Salaried Masses*, which has the German title *die Angestellten* (the employees), charts the rise in a revolutionary new form of employment that was transforming Berlin after the First World War. This work, as discussed by Kracauer, involves operating the analog wire technologies and early computing machines that were beginning to reform economic and governmental processes. The resulting (machine-mediated) human interaction during newly coordinated 'office hours' contributed to the networks of information technologies that have been further established ever since. The new employment opportunities lead to large scale migration to major cities. The proliferation of disposable incomes created service, culture and leisure industries, and with this further employment prospects. To introduce his readers to the defining trait of this new form of work, Kracauer questions a young secretary who is on a commuter train on her way home to a Berlin suburb. She cannot understand why anyone might be interested in her mundane daily routine. Yet it is precisely this feeling of insignificance, ordinariness and everydayness that Kracauer's study presents as the absolute novelty and the unobserved by-product of the proliferation of wire-technology. In contrast to the strenuous work in the field, factory or household service that salaried employment was replacing, the daily routine of successive telephone calls, the typing and processing of faceless reports and data leave the impression that *nothing is done at work*, or at least, nothing worth thinking or talking about outside of the office. Despite turning up for work, these employees remain curiously absent. The one *tangible* product of the time spent operating a machine is an expendable income, payment, which can, no *must*, be spent in newly established department stores, cinemas and cafés during 'free time' after work.[1]

By recording the view the secretary has of herself and of her activities while sitting on a train that is leading her through a rapidly changing world, Kracauer begins to document the façade of a lived and experienced ordinariness that is maintained and even strengthened by the introduction of new technologies, rather than being challenged by their often extraordinary capabilities. His compelling snapshot of Berlin in the late 1920s shows us the inconspicuous veneer of normality behind which tremendous (and catastrophic) societal forces were gaining momentum. Cinema, shopping mall, internet and our own consumerist lifestyles are rooted in this production of salaried ordinariness, as much as Hitler, the atom bomb and globalised industrial production are. What Kracauer captures in his portrayal of Berlin is the nascent state of our own increasingly fractured, unequal, dehumanised, polluted, but yet totally ordinary world. This is a world that is collectively produced by billions of human hands that are radically out of touch with the end product of their activities: an inverted utopia, a world which, as a product of our growing mastery, is running out of control.

Today, as new 'worlds' are rapidly configuring themselves around ever 'smarter' technologies, Kracauer's study has gained renewed significance. Amid the voices that diagnose the unprecedented nature of the challenges awaiting us in the digital age, it is worth recalling the famous diagnosis that Walter Benjamin offered in 1933, just days after fleeing Berlin to Paris on account of Hitler's election. For Benjamin, who is partly responding to Kracauer's writings here, the desire to turn to the future and pretend that the world is 'starting from scratch' is part of what he calls a 'poverty of experience' which takes hold as a result of rapid technological progress.[2] This poverty consists of an inability to comprehend and communicate the reality to which technological change exposes us as individuals. As a consequence of this poverty, Benjamin concludes, 'nature and technology, primitiveness and comfort, have completely merged'. A reality is established in which people flee their incomprehension by retreating ever further into the anaesthetic world of machines:

> To people who have grown weary of the endless complications of everyday living and to whom the purpose of existence seems to have been reduced to the most distant vanishing point on an endless horizon, it must come as a tremendous relief to find a way of life in which everything is solved in the simplest and most comfortable way, in which a car is no heavier than a straw hat and the fruit on the tree becomes round as quickly as a hot-air balloon.[3]

Anders's inverted utopias pick up this Benjaminian problematic to introduce us to a further aspect of this poverty of experience, one that Bernard Stiegler has recently addressed at length by diagnosing a general crisis of '*savoir vivre*'.[4] This crisis manifests itself in our inability to live *without* the assistance of the increasingly capable technological objects that seem to simplify life and make things easier for us. *The Salaried Masses* reveals a strange parallel between the 2010s and the late 1920s. Although the effects of digital technologies and their current, anticipated or potential capabilities are the subject of sustained media speculation and academic reflection, these same technologies are treated as utterly ordinary and inconspicuous as soon as they find their way into our lives. Digital devices, it seems, are 'simply there'. They are encountered, to reiterate Anders's observation about analog machines, 'primarily as necessary, desirable, superfluous, affordable or unaffordable' things that we can procure in shops, put in our pockets and use to work, keep in touch and stay connected to news feeds.[5] We seem to question their existence, their wider effects and the way they shape our thoughts, feelings and behaviour as little as the secretary in Kracauer's study questions the 'normality' of the phone she answers in return for a regular income.

The feeling of ordinariness Kracauer isolates leads us to the heart of the problematic that Anders's 'Promethean conception' of technology exposes. Because a relationship to technological artifice comes naturally to us, the trickery of machines slips from sight the more it is incorporated into our intuitive sense of self. The secretary, to return to Kracauer's opening image, does not contemplate or comprehend (and cannot fully do so) that her being on call in the office in the first place is already the mark that the parameters of existence have undergone fundamental reconfigurations. As a consequence, therefore, the very changes technology effects happen in a perceptual vacuum and remain under-scrutinised. This 'invisibility' is protected, as Kracauer elaborates while reflecting on the life of employees, because it is deemed 'too ordinary' to merit critical study. More significantly still, those in a position to question the ordinariness of the machine interactions in question are themselves employees (at universities, newspapers and cultural institutions). As such, they rely on the very machines at the heart of these developments and are bound by the precarious independence, by the mundane hopes, anxieties and aspirations that a regular, disposable income sets in motion. 'Even radical intellectuals', Kracauer concludes, 'do not easily get behind the exoticism of a commonplace existence,' for they live it themselves (p. 29). Like Edgar Allen Poe's '"Letter to Her Majesty,"' the transformative power of salaried and paid employment is missed and remains unchecked because it is so openly on display.

Today we are in the midst of a shift in the technological environment that far exceeds the effects and forces unleashed by the widespread introduction of 'analog' technologies. Amid the fears of falling out of sync with the online environments that new information technologies enable, and in view of the pressure to build up online profiles and the scramble to develop forms of academic research and outputs that harness the power of digital technologies, Kracauer's opening diagnosis about the curious ordinariness and under-scrutinised nature of technological change still seems highly accurate: 'Powerful forces are at work anxious to prevent anyone noticing anything here' (p. 29).

To introduce us to these 'powerful forces', Kracauer leads us onto the commuter train. The secretary's impression that she is doing nothing extraordinary at work is presented as being symptomatic of a collective blindness: 'Hundreds of thousands of salaried employees throng the streets of Berlin daily yet their life is more unknown than that of the primitive tribes at whose habits those same employees marvel in films' (p. 29). With this image we are introduced to the peculiar discrepancy that begins to unfold in the 1920s and that persists to this day. Although new technologies continually transform and profoundly affect the realities of work, entertainment, consumption and human interaction, we marvel at the adventurously exotic, strange and fictional worlds these technologies invite us to watch, instead of contemplating the strangeness, novelty and exoticism of our own. By the late 1920s, it seems the 'force field of destructive torrents' with which technological progress had exposed a newly 'fragile' and 'tiny' human body during the First World War, had mellowed into an unsettlingly comfortable, anaesthetic breeze.[6] Kracauer shows us that the joyful surrender to 'home comforts', after work entertainment and shopping, also stirs up new forms of hypersensitivity towards what are felt to be painful, and inescapably personal shortcomings and flaws. Adverts for beauty products, dinner parties, rejuvenation elixirs and pharmaceutical supplements begin to openly cultivate and exhibit new social pressures and private anxieties with which employees have to contend. Alongside the analog information technologies of the 1920s new fragilities and worries emerge, ones that employees take great pains to conceal.

The Salaried Masses hence teases out what Antonin Artaud would have described as 'a latent undercurrent of cruelty' building up momentum beneath the inconspicuous fabric of everyday life.[7] It reveals the emotional wounds that employees were beginning to suffer by assembling a 'mosaic from single observations' of individually inconspicuous procedures, behaviours and attitudes into a narrative that traces the contours of the depersonalising, dehumanising, humiliating and now ubiquitous forces of abstraction that new information technologies were beginning to mobilise. The phenomena Kracauer isolates in order to impress on his readers the strangeness of this seemingly ordinary world are strikingly familiar. He begins by dissecting the language of job adverts and rejection letters, and analyses how highly overqualified candidates compete for work with 'handsome weapons in the form of certificates and diplomas' (p. 34). In an unnamed corporation, 'machine-girls' take us on their smoking break because the 'Powers punch-card book-making machines' they operate need a 'short break for ventilation' (p. 41). We move on to the wages and personnel office where 'all kinds of pre-printed forms' are fed into machines, as the work force is continually restructured in response to further technological solutions (p. 42). This large scale depersonalisation of individuals through certificates, mechanised bureaucracy and HR departments has its flip side in the hypersensitivity of employees who are worried that they will not make the grade or might lose their job.

We are introduced to the uniform, orchestrated smiles of sales assistants, and their efforts to appear young, upbeat and energetic. 'For fear of being withdrawn from use as obsolete, ladies and gentlemen dye their hair, while forty-year-olds take up sports to keep slim' (p. 39). A magazine headline asks '"How can I become beautiful?" (p. 39) and with this Kracauer contemplates the manner in which 'fashion and the economy work hand in hand' (p. 39), one stoking the fire that the other sustains. The complementary nature of consumption and a competitive labour market is most evident when employees return home after work. For 'home', as Kracauer puts it with indelible wit, is 'not just a lodging' (p. 88). 'Home', rather, 'is an everyday existence outlined by the advertisements in magazines.' These show 'pens; Kohinoor pencils; haemorrhoids; hair loss; beds; crepe soles; white teeth; rejuvenation elixirs; selling coffee to friends; dictaphones; writer's cramp; trembling, especially in the presence of others; quality pianos on weekly instalments; and so on' (p. 88). Like our internet search history today, the adverts, corresponding possessions and social interactions that Kracauer records act as a catalogue of the private insecurities, hopes and desires that are cultivated by the very things and products that so inescapably promise relief. Because they help diffract, direct and fix attention, 'magazine adverts', to fit a memorable sentence by Kracauer to the current image, 'show the very life that employees cannot settle into'.[8] Viewed from the perspective Kracauer opens up, the objects we own and use become representative of the intangible forces that are animated while we interact with machines. The 'enjoyment of entertainment' (*Zerstreuung* in German) also mobilises forces that remove us from the ground we stand on. These forces can 'scatter us' to use the other meaning of the word *Zerstreuung* that Kracauer repeatedly plays on; they have the capacity to pull our lives apart.

As in 1920s Berlin, therefore, perhaps the most exotic aspect of the adoption of new (digital) technologies today is how inconspicuous these remain *despite* their obvious transformative effects. In principle, most computer, mobile device or smartphone users would describe their activities as being totally ordinary. It seems likely that we would intuitively respond to the question 'what do you do while waiting for the train, during your lunch break, or while surfing the web' with a shrug of the shoulder that indicates that we are doing *nothing*, thus mirroring the attitude Kracauer's study seeks to portray. Günther Anders has scrutinised the effects of this inconspicuous (in)activity, this '*perceptual doing nothing*' while interacting with machines at work or during hours of machine mediated leisure, perhaps more extensively, pragmatically and certainly more polemically than most twentieth-century thinkers. Anders concludes that our sense of ordinariness means that complex technological objects can no longer be considered observable 'phenomena' in the traditional sense: 'If one understands the word "phenomenon" as meaning "something that shows itself," as Heidegger does, then the machines [of today] are no longer phenomena.' They look ordinary, '*like nothing*', '*they seem much less than they are*'. As part of their retreat into invisibility they also conceal the part they play in shaping our lives and the world we are part of.[9] In doing so, they blind us to the potential they have to create conditions in which we begin to devalue each other and ourselves. Far from being an exercise in polemic rhetoric, then, Anders's adoption of exaggeration as a method is rooted in the paradoxical phenomenology of self-concealing things. Because it presents us with a sense of ordinariness and normalcy, as Anders argues in his 'Theses for the Atomic Age', 'perception is a "false witness"': 'Today, it is not in the wide land of imagination that escapists like to hide, but in the ivory tower of perception.'[10]

In *A Visit to Hades*, we are offered a longer explanation of the task to which philosophical exaggeration responds. I will cite this important passage in full:

> Our faculty of perception is too limited to enable us to comprehend the state of the world today. It is too short sighted to show us the enormous, or rather, the monstrous dimensions of our deeds, because it continually transforms abject monstrosity into something that is inconspicuously ordinary. As nonsensical as this may sound, our perception has become a variety of phantasy and imagination. Anyone who believes that the world is the way that he perceives it is a fantasist.

The critical task that Anders proposes as technological objects increasingly put us out of touch with our acts consists of resisting and challenging the sense of ordinariness that our bodily senses, our perception, intuition and our highly selective sense of compassion lure us into. To do so, he continues, 'we must begin to use our imagination as a corrective force':

> The task of imagination required today departs from what imagination has meant up to now. [...] It no longer means the act of imagining fictitious things or picturing imaginary creatures. [...] Quite the opposite: the task consists in mobilising our imagination as a way of approaching the truly fantastical reality of the world today, so we can grasp this reality in an appropriate manner. In brief: *Imagination* [...] *needs to become an empirical method, an organ of perception for that which is truly enormous.* [...] Imagination, like a telescope, does not make our organs of perception superfluous. It is only when we use it that we give our perception a proper chance to see and comprehend.[11]

By assembling individually inconspicuous observations and interviews into a narrative, Kracauer's study arrives at an exaggerated perspective that anticipates the diagnoses offered by Anders's inverted utopias. What he begins to identify, therefore, is that the comfortable, anaesthetising sense of ordinariness that salaried work can instil contributes to the concealment of the large scale effects of technological change, which have fundamental repercussions for individual workers, and the planet as a whole. As we have seen, the tangible emotional issues *The Salaried Masses* introduces us to are the anxiety to belong, the fear of losing (or not finding) employment, and the resulting hypersensitivity, negative self-scrutiny and drive for self-improvement. Put otherwise, and in the words of Anders's own reflection on Berlin in a 1931 essay, employees are too preoccupied with 'the narrow confines [. . .] of their own business' to scrutinise the technological objects that begin to populate their lives.[12] This self-centredness anticipates one of the key theses that 'On Promethean Shame' advances, namely that we can begin to feel singled out, isolated and inadequate in view of the worlds mediated by information technologies, or even, in response to the superior capabilities of our technological objects, including their ability to command the attention of those dear to us.[13]

Even if they might be hard to observe and quantify, the affective states that accompany our interaction and competition with machines 'show themselves' (albeit primarily as feelings of discomfort, loneliness and private concern). As such, they still belong to the realm of traditional phenomenology, which Anders deems too limited to apprehend the often destructive modes of production and social realities that configure themselves around complex technological objects. In fact, the very hypersensitivity that throws us back onto ourselves and our own limitations, Anders suggests, is one of the traits that turns our attention away from the extreme insensitivities, carelessness and absentmindedness that is cultivated by our increasingly capable machines. In the remainder of this book I will respond to Anders's suggestion that '*imagination needs to become an empirical method*' and travel into the unobserved world of mundane machine interaction.

If the key social phenomenon accompanying the shift to analog technology was the proliferation of 'salaried employment', then a comparable development today is the rise of what I call 'unsalaried' work in allusion to *The Salaried Masses*.[14] Viewed through the perspective Anders opens up, this term does not only seek to capture the widely studied proliferation of forms of unpaid 'intellectual labour', the emerging 'attention economies', the automation of service industries and the neoliberal exploitation that digital technologies enable.[15] It names *the labour we all conduct towards our own obsolescence*. This form of unpaid activity consists of the effort we exert to combat our 'own limitations only to limit ourselves in new ways'.[16]

As such, the obsolescence of individual workers or whole lines of work are merely the surface phenomena of a more fundamental form of obsolescence. The recent work of Bernard Stiegler succinctly introduces the obsolescence that becomes ingrained in human subjectivity as we turn to ever more powerful technological objects. In 'Anamnēsis and Hypomnēsis', Stiegler elaborates that the more we turn to memory aids, our mobile phone for example, the more we also fail to exercise our own memory. This has been outsourced to the phone, which means that the price we pay for our technologically enhanced memory is the loss of our own. 'Mislaying a mobile phone' as Stiegler puts it, 'is equivalent to losing track of the numbers of the people one is in contact with and realising that they are no longer in one's own memory, but in the device itself.' This 'knowledge that escapes us', as Stiegler elaborates,

> seems to induce 'human obsolescence', which finds itself thereby more and more deprived, as if hollowed out from the inside. Thus, the more cars become perfected – the navigation system that assists the driver today in his driving will replace him completely tomorrow: he will control the vehicle from a distance through a system of automatic driving – the less we know how to drive. [...] The more we delegate the assumption of the series of little tasks which make up the framework of our lives to devices and modern industrial services, the more futile we become: the more we lose not only our know-how (*savoir-fair*) but also our *savoir-vivre* – and with this the little pleasures that make life worth living. We end up only fit to consume indiscriminately, without the pleasures that knowledge alone can provide – as if we were impotent.[17]

Stiegler's provocative reflection highlights the two antithetical forces that Anders isolates in his work: the more perfect machines become, the less perfect humans need to be and the more limited they begin to look; the more capable machines are, the more they have the potential to produce incapacitated humans. Put otherwise, the Promethean tendency to combat our limitations also has the tendency to cultivate and arrest us in the impotent, 'Epimethean' state of lacking qualities, because captivating technologies relieve us of the effort of acquiring the intangible, immaterial skills of entertaining ourselves, finding composure in social situations and acquiring the 'life skills' (*savoir-vivre*) that Stiegler sees under threat.

Stiegler's reflections on this outsourcing of knowledge are oriented by Gilles Deleuze's highly influential discussion of the 'Societies of Control' that begin to emerge after the Second World War.[18] The short parable with which Anders opens 'The World as Phantom and Matrix' – the substantive 1956 essay on TV and visual culture that follows on from 'On Promethean Shame' in *The Obsolescence of Human Beings 1* – perfectly anticipates the problematic that Deleuze, and in turn Stiegler, present us with:

> Because the king did not like how his son – straying from the roads under the king's control – was cutting across country in order to form his own image of the world, he gave him a horse and cart. 'Now you no longer need to go on foot', were the king's words; 'now you are no longer allowed to do so', was their meaning; 'now you cannot do so anymore' was their result.[19]

This story succinctly introduces one of the main components of Anders's notions of human obsolescence. The prince pays for his more comfortable journey with the free use of the legs he stands on. The king controls his son, not with prescriptions, but by making his life easier and more convenient. In accepting his father's gift, the son actively participates in his own subjection, for the more he relies on the cart, the less he knows how to live without it, the more willingly and unwittingly he plays by his father's rules. The increasing potential that technological objects have to make us *yield control over our lives* by making it easier is the key tension I will discuss in the remainder of this book. For Anders, the emergence of 'control societies' is the symptom of a still more fundamental form of obsolescence: the obsolescence of the faculties and bodily registers that supposedly make us human. What escapes us when we turn to complex machines is not just our *savoir vivre*, but also our ability to deliberate our acts and our bodily apprehension of their outcome, that is, our joy, fear and shame, but also the hate, malice and selfish desires that we are accustomed to associate with the suffering that humans are capable of inflicting on each other. Anders challenges us to see that it is precisely the realm of human feeling and the role it plays in mediating our responsibility and responsiveness towards others that is bypassed, the more we outsource our lives onto machines.

NOTES

1. See Siegfried Kracauer, *The Salaried Masses: Duty and Distraction in Weimar Germany*, trans. Quintin Hoare (London and New York: Verso, 1998). All references are from this edition and are given in the text. For a discussion of the orchestrated 'unfreedom', which, as Kracauer suggests, 'free time' constitutes, see Theodor Adorno, 'Free Time', *The Culture Industry: Selected Essays on Mass Culture* (London: Routledge, 1991), 187–97.
2. See Walter Benjamin, 'Poverty and Experience', *Selected Writings Volume 2: 1927–1934*, trans. Rodney Livingstone (Cambridge, MA: Harvard University Press, 1999), 732.
3. Benjamin, 'Poverty and Experience', 731–5, 735.
4. See Bernard Stiegler, *What Makes Life Worth Living: On Pharmacology*, trans. Daniel Ross (Cambridge: Polity, 2013). I return to this point in detail below.
5. Anders, 'On Promethean Shame', (ch1) 33*.

6. See Benjamin, 'Poverty and Experience', 732: 'A generation that had gone to school in horse-drawn streetcars now stood in the open air, amid a landscape in which nothing was the same except the clouds and, at its centre, in a force field of destructive torrents and explosions, the tiny, fragile human body.'

7. See Antonin Artaud, *The Theatre and its Double*, trans. Victor Corti (Richmond: Oneworld Classics, 2010), 20f.

8. I am here referring to the following passage from Kracauer's 'Photography': 'In the illustrated magazines, people see the very world that the illustrated magazines prevent them from perceiving. [...] Never before has a period known so little about itself. In the hands of the ruling society, the invention of illustrated magazines is one of the most powerful means of organising a strike against understanding' (p. 58). See Siegfried Kracauer, 'Photography', *The Mass Ornament: Weimar Essays*, trans. Thomas Levin (Cambridge, MA: Harvard University Press, 1995), 47–64.

9. Anders, *die Antiquiertheit 2*, 34–5. The illustration continues as follows: 'Technological objects [and the networks they form] are mute: their appearance no longer discloses anything about their characteristics, traits and background. Granted, the expression "mute" is not entirely precise, for it cannot be denied that technological objects can be perceived. Rather, they remain unrecognisable and unknowable even though they are in some form perceivable. They shelter behind an appearance that has nothing to do with their nature, they appear to be less than they are.'

10. Anders, 'Theses for the Atomic Age', *The Massachusetts Review* 3.3 (Spring 1962), 493–505, 497.

11. Anders, *Besuch im Hades* (München: C.H. Beck, 1996), 39–40.

12. Günther Anders, 'Der Verwüstete Mensch', *Mensch ohne Welt* (München: C.H. Beck, 1993), 3–31, 3.

13. The last point, in particular, seems to be one of the principal domains of 'Promethean Shame' today. On this point see Sherry Turkle's discussion of the behaviour of teenagers who have grown up in an environment in which parents were often preoccupied with their phones: 'From the youngest ages, these teenagers have associated technology with shared attention. Phones, before they become an essential element in a child's life, were the competition, one that children didn't necessarily feel they could best' (Turkle, *Alone Together*, 267).

14. In German I call this '*unangestellte Arbeit*', that is, 'unemployed work', but I opted for 'unsalaried work' to keep the allusion to Kracauer's study. This work, as we will see, is unpaid and takes place as soon as we interact with a networked device. As 'unsalaried work' might imply forms of payment other than a salary, I tried to avoid this term in my discussion and use 'unpaid work' instead.

15. I am using these terms, which I will specify further below, with reference to Patrick Crogan and Samuel Kinsley, 'Paying Attention: Towards a Critique of the Attention Economy', *Culture Machine* 13 (2012), 1–29.

16. See Anders, 'On Promethean Shame', (ch1) 28*.

17. See Bernard Stiegler, 'Anamnēsis and Hypomnēsis: The Memories of Deisre', *Technicity*, ed. Louis Armand and Arthur Bradley, trans. François-Xavier Gleyzon (Prague: Litteraria Pragensia, 2006), 15–41, 17f.

18. In the context of this discussion, it is worth recalling that Deleuze presents salaried employment as a defining factor contributing to the emergence of Control Societies. See Gilles Deleuze, 'Postscript on the Societies of Control', *Negotiations*, trans. Martin Joughin (New York: Columbia University Press, 1995), 177–82, 179.

19. Anders, *die Antiquiertheit des Menschen 1*, 97.

Chapter 3

The Unsalaried Masses

Working towards a 'World without Us'

For they do not know what they are doing.

To begin our excursion into the world of 'unsalaried work', we can follow Kracauer's lead and step onto a commuter train. Kracauer would find aspects of the scene highly familiar, as the train passengers are likely to be employees who are on their way home or travelling to work. New sounds, however, have joined the rustle of newspapers and the muted conversations that would have already been part of the ambient noise in 1920s Berlin. The hum of headphones, the sound of key strokes, phone alerts and the public performance of apparently confidential conversations are an indication that these workers also live an additional, 'unpaid' routine. As banal, insignificant and inconsequential as this may sound, this form of 'unpaid work' consists of mere machine interaction and personal dependence, yet this activity may cost a large proportion of salaried employees their employment (and much more) in the future. For today, the interaction with networked machines, and via these with other humans, invisibly generates and accumulates data and meta-data and translates human spontaneity and patterns of behaviour into the format that 'smart' devices require in order to assist, helpfully mimic and conveniently outperform their users.

'Unsalaried work', as portrayed here, is a further evolution of an activity that Anders introduces as 'unpaid work from home' (*unbezahlte Heimarbeit*) in 'The World as Phantom and as Matrix'.[1] This peculiar form of work, Anders suggests, consists of the *mere use* of tele-technologies that stimulate our most receptive and seemingly 'immediate' senses: our vision and hearing. Through the comparatively passive act of habitually consuming audiovisual media content, Anders contemplates, we *actively produce* a new type of subjectivity, a new type of human being even, who understands its own existence through the 'mass' mentalities and attitudes configuring themselves around the interaction with these captivating technologies. The content engaged with during the hours of unregistered 'work' is irrelevant. What counts, rather, is the act of *cultivating a dependency*, as this creates an environment in which we begin to inhabit the unreal but yet immediate phantom worlds that captivating tele-technologies present us with in a liquid, pre-digested form. In order to become receptive for television, the owners of television sets must actively work at adjusting their habits and behaviours; they must stay at home and isolate themselves from others in front of television sets in order to receive a transmission that is enjoyed alongside millions of others. Anders presents the 'mass-solipsism' that the advent of television heralds as the beginning of a new form of 'cottage industry', which is animated by the unpaid efforts each viewer exerts to tune into the transmission: 'These homeworkers conduct the work required to transform themselves into mass human beings through the consumption of mass content. Their work consists of *leisure*.'[2]

This unpaid employment is 'fully paradoxical' as Anders elaborates, because 'the homeworkers need to pay for their efforts (the television set, often also the programmes) instead of being paid'. Their efforts amount to buying the right to passively expose themselves to advertisers and politicians, whose power and influence is contingent on a receptive audience. The absurd situation Anders contemplated in 1956, before television sets had even been widely introduced, has escalated in the context of digital culture. As is well known, services such as *Google*, *Facebook*, *Twitter* and *YouTube* rely almost exclusively on user-generated content. While interacting with online services, as William Gibson puts it in the *New York Times*, we routinely act as '*unpaid content providers*' whose time, attention and effort (i.e., whose 'work') *creates and guarantees the stock market value of these companies*.[3] The problematic Gibson isolates, however, is merely the small visible tip, the economic dimension, of this form of unpaid work. As Jaron Lanier has discussed at length, mere interaction with digital devices produces data sets that can be 'mined' for algorithmic applications. Lanier forcefully argues that each click we make online should therefore lead to an automatic 'nanopayment' that would acknowledge our 'economic ownership' of the data our machine interaction generates.[4] In the unlikely event that such payments became widely adopted, this would only offer small consolation. For, whether we like it or not, the time we spend online contributes to our own obsolescence in more than just an economic sense. The data we generate is the capital of the digital age, not only because it is enabling the automation of seemingly distinct human capabilities, thus potentially rendering human contributions obsolete or at least economically unviable, but also because it invites a more fundamental and destructive form of obsolescence – the obsolescence of 'our heart', as Anders calls it, that is, the obsolescence of 'our inhibitions, fears, worries, regrets'.[5] The economic dimensions of obsolescence that Anders contemplates in his extensive discussions of the machine assisted division of labour are focused onto the emotional detachment between ourselves and the actions we are implicated with. This detachment is amplified the more complex and autonomous technological objects become. Whereas Kracauer focused on the emotional wounds employees suffer on account of a labour market in which they are replaceable, Anders no longer focuses on the phenomenological realm of observable emotional repercussions of being an employee, but on the untraceable and intangible chains of abstraction that the work of hundreds, yes even thousands of employees add up to. For these forces of abstraction are the Promethean chains that Anders seeks to contemplate, as these are the chains that bind us to the deceptive feeling of ordinariness that our senses present us with.

On the basis of my opening engagement with *The Salaried Masses*, such a chain of abstraction severing the bond between the employee's everyday concerns and the cumulative effect of the efforts of hundreds of employees might look like this: information compiled elsewhere is fed into 'a Powers punch-card bookmaking machine', to stay with Kracauer's example. This machine is used to evaluate the efficiency of a specific work process, and how it might be improved. If a technological means of increasing efficiency is available, Human Resources is informed. Files are consulted, previously typed reports are read and a superfluous worker is isolated. The standard pre-formulated letter is prepared in the legal department. It is posted by someone else, or even by a machine. The cumulative effect of these mundane partial steps is that a life is thrown out of the fold of normality, without anyone being in direct contact with the ultimate outcome of their mundane activity. Because various technological registers act as mediators that maintain and generate a distance between the employees and the outcome of their collective work, salaried employment remains inconspicuous. The fate of the sacked individual is destined to remain abstract, and, unless one knows this worker personally it is not part of 'the narrow confines of one's own business'. Even when an employee is aware of the immediate consequence of their activity, the sacked worker's fate cannot be prevented, for the employee is merely doing a job that another would have done if one had refused to do so. '*The division of labour*', as Anders concludes, '*is always also a division of conscience.*' While working, 'nothing is more successful, than the vulgar generosity with which an excusing inner voice grants us absolution for our acts [...]: "If you don't work, *someone else* will do so in your place, and the outcome will not change in the slightest."'[6] As 'On Promethean Shame' reminds us, 'viewed from the perspective of institutions, the economy, the leisure industry, politics, and warfare [...], "*spare men*" are always available.'[7]

This inevitably simplified, schematic description of the steps that effect the redundancy of a worker merely seeks to illustrate the equation with which Anders describes technological progress: 'The tendency without which no machine would be a machine, is the aim of concentrating a maximal degree of efficiency and power with the lowest possible investment of human effort. This is the idea inherent to technology.'[8] At first glance, this may simply seem to be an elaborate way of saying that machines make things 'easier'. Their mediation makes it easier to work but also easier to find distraction and entertainment. What Anders introduces us to here, however, is what he calls 'the oligarchic principle of technology'. This principle seeks to describe how the salaried or leisurely interaction with a machine leads to an *automatic* concentration of power (and also wealth). For machines reduce effort in two distinct ways. *They increase the productivity of the worker, and simultaneously create (social) conditions in which the rationalisation and influencing of individual behaviour is met with less and less resistance.*[9] It is here that we can begin to understand why Anders insists that the mere existence of a machine amounts to its use, for it either prescribes particular relations and modes of interaction or it '"sets the ground" for such relations to unfold'.[10] The emerging technological networks Kracauer describes were *not directly or exclusively* invented in order to find applications in the office. Rather, it is their very existence that illuminates potential applications and unanticipated opportunities to reduce a workforce or automate a specific process. The consequence of this 'oligarchic principle' is that power begins to be concentrated into the hands of those who history will reveal as having had an understanding of how to effectively mobilise technological innovations, or see their innovation, application or hardware widely adopted. The very fact that technological progress has unfolded through the creation of a vast inventory of unsuccessful technological objects that look like the product of misjudgements or ludicrous speculation in retrospect, is in itself an indication that our scheming Promethean foresight is far too limited to give us an indication of what objects and applications might succeed (and what they might be used for).

In the *Decadence of Industrial Societies*, Stiegler cites the slogan of the Xerox Research Centre, which perfectly captures the strategy that is adopted to combat our limited ability to anticipate the future: '*The best way to predict the future is to invent it.*'[11] Stiegler's discussion of control societies goes on to highlight how this mantra has become a strategy, not just for the business world, but also one adopted by state power. In doing so, he succinctly illustrates the consequence of the oligarchic principle. The more technology mediates our lives, the more market leaders and investors in infrastructure are also in a position to shape the future because they have the resources and market shares to promote the technological solutions around which future lives will revolve. It is worth noting that the uses to which these future objects will be put, does not have to be determined or predicted as any use will secure the future value of the company. By highlighting the oligarchic principle, Anders does not seek to chastise technological progress as the effect of evil machinations. What he describes, rather, are the conditions under which such machinations become increasingly possible, take less 'effort' and can be met with less resistance as technological infrastructures are ever more integrated not only into economic processes, but crucially also into our sense of self. The obsolescence of individual lines of work that may ensue from technological progress is likewise not the product of a malignant desire to create unemployed workers, but a consequence of the desire to attain a workflow that does not require any employees.[12] As is corroborated by statistical data, for instance by 'the Great Decoupling' of the American economy diagnosed by Erik Brynjolfsson and Andrew McAfee, the oligarchic principle is creating an environment in which there is an ever widening gap between the winners who profit from the emerging networks and technological infrastructures, and who can ever more effectively ensure that they will continue to do so, and the losers whose efforts are no longer required as the digital revolution unfolds and who are increasingly locked out of the labour market.[13]

The paradox that Anders highlights in line with Kracauer's diagnosis is that salaried employees only become employees because they are needed to operate machines. As is explained in 'On Promethean Shame', 'by extinguish[ing] what they are *as* acting individuals', by '*transform[ing]* their actions into *mere processes*' employees contribute to the development and indispensability of technological and organisational infrastructures that expose them to the risk of becoming obsolete once a more capable or powerful machine becomes available.[14] For the more our lives, production processes and intellectual and cognitive activities configure themselves around machines, the less humans are needed as machines become more capable. Put otherwise, and more succinctly: *employees are working themselves out of work.* The more efficient they become because of technological advancement and the more quantifiable their patterns of consumption become, the more human employees become dispensable. This is a first indication of why Anders suggests that 'today the true robots are not the pretend-humans (*Scheinmenschen*) that are assembled from dead components and actualised in "computing machines," but the machine-parts that living humans function as'.[15]

The schematic outlined so far has described the comparatively harmless fate of individual employees. It has deployed the term 'obsolescence' in its conventional sense, as meaning that a human being, or particular technological device or idea has become antiquated and is no longer competitive. This, however, is only a side product of the form of obsolescence that Anders seems to primarily have in mind when using the term. To outline what is ultimately at stake when thinking the 'unsalaried work' of interacting with a networked device, I turn to a brief discussion of this more fundamental notion of obsolescence by returning to Anders's analysis of the most powerful and destructive technological objects in existence: nuclear weapons.

An infernal parallel begins to open if we read Harry Truman's press release from the 6 August 1945 alongside Kracauer's caricature of the secretary's impression that she is doing 'nothing' at work. For, as Harry Truman explains to the world public, the bomb deployed against Hiroshima was produced by thousands of employees:

> Employment during peak construction numbered 125,000 and over 65,000 individuals are even now engaged in operating the plants. Many have worked there for two and a half years. *Few know what they have been producing*. They see great quantities of material going in and *they see nothing coming out* of these plants, for the physical size of the explosive charge is exceedingly small (my emphases).[16]

While celebrating the organisational efficiency of the US state apparatus and threatening Japan with a physically very small but immensely powerful explosive device, Truman succinctly introduces the basic product that machine mediated employment creates according to Anders's writings: '*nothing*', that is, nothing to worry about. Ultimately, Truman's threat is backed up by the successfully co-ordinated and 'secret' work of thousands of employees. Truman's true source of power is not the atom bomb, but *the ignorance and incomprehension of its thousands of producers, and even the ignorance of those involved in its deployment.*[17] I have highlighted the word 'secret' above, because secrecy is only needed to protect the exclusive ownership of the *assembled end product*, and the knowledge of how to assemble all the required individual components. A prime function of secrecy, as Joseph Masco has shown, is the perfect pretext it offers for the consolidation of state power.[18] In all other respects, however, secrecy is achieved by default. The complexity of the bomb itself, the convoluted production, administrative, logistical and scientific processes leading to its existence make it impossible for one of the 125,000 individual workers to comprehend the part they played in producing a device that instantaneously killed over 110,000 people. The device severs the link between the outcome and the thousands of (in)directly implicated humans. Employees are also constantly aware of their replaceability if they were to opt out. The bomb thus actively serves to sanitise power and police a way of life, not least by impressing on each individual how powerless they are to do anything about its existence. Its power is simultaneously aesthetic and anaesthetic.

For Anders, then, this extreme compartmentalisation, automation and division of labour begins to instil a condition of *amorality* that takes hold because employees cannot possibly retrace the ultimate effects of their own activities. The point is not that employees 'should know what they are doing', but that the complexity of the production processes and one's own replaceable and seemingly negligible contribution prevent such knowledge, and more importantly, a *feeling* of scruple, in advance. The very existence of the bomb is testament that we have been propelled into a sphere 'beyond morality and immorality' because 'the division of labour prevents the [participant] so completely from having clear insight into the productive process, that the lack of conscience we must ascribe to him is no longer an individual moral deficiency'.[19] It would be 'inappropriate', Anders suggests, to call a nuclear explosion a human '"act" (*Handlung*)', because 'the chain of events leading up to the explosion is composed of so many links, the process has involved so many different agencies, so many intermediate steps and partial actions, none of which is the crucial one, that in the end no one can be regarded as the agent. Everyone has a good conscience, because no conscience was required at any point.'[20] The situation is *amoral* (not merely immoral) because those linked into chains of abstraction (i.e., all of us) remain '*excluded from the possibility of having a conscience*' in the first place.[21] Diligence and conscientiousness at work are potentially highly incommensurate with the kinds of moral cognition that mediate more 'direct' human interaction. At work, our moral sensibilities remain oriented by the concrete human interactions and responsibilities that form part of the everyday labour process. They cannot at the same time intuit the ultimate object, effects (and possible effects) to which the product of their work contributes now or in the future (not least because future uses cannot be anticipated). 'The specialised worker', Anders concludes, 'is not conscious of the fact that the conscientious efforts of a number of specialists can add up to the most monstrous lack of conscience'.[22]

The statement about Hiroshima issued on the 6 August 1945 perfectly illustrates the removal of the human that Anders's notion of obsolescence seeks to capture. Truman, despite praising the efforts of the scientists and employees who created the bomb and the military personnel who deployed it, presents the *atom bomb* as the true agent. 'The greatest marvel of all', as Truman ends his account of how the device came into existence, is that 'the brain child of many men *came forth in physical shape* and performed as *it* was supposed to do'. This rhetoric perfectly illustrates the bleak and disarming conclusion that Anders arrives at while presenting the atom bomb as the exemplary technological object that illuminates the tendencies of all others: despite being the direct product of human activity, despite being possible only because of the cumulative effects of the individually inconspicuous activities of employees and consumers, apparently '*inhuman deeds*' are '*deeds without humans*' and are registered as such.[23] Who would you rather shake the hand of, Anders asks us to consider, a bomber pilot returning from a mission or a man who has just brutally strangled someone with his bare hands?

The atom bomb, as the most powerful technological object with an incomprehensible, awe-inspiring explosive power, perfectly embodies the 'oligarchic principle' that Anders ascribes to all complex technological objects. It concentrates 'a maximal degree of efficiency and power with the lowest possible investment of human effort'. It is a 'better', more reliable, cheaper, more efficient, less inhibited killer than an army carrying guns. The administration of deadly force and the work required to maintain a constant threat now only requires an army of specialised (morally) good workers (who engage with machines), not an army of vulnerable 'killers', who would have to 'work through' their actions, be motivated to kill, and whose loss would be mourned, thus possibly destabilising the administering power.[24] 'Nothing would be more short-sighted', as Anders elaborates, 'than to consider the possibility of our own liquidation the arbitrary by-product of a few special devices, nuclear weapons for example.' The atom bomb, rather, *openly portrays* 'the principle' all complex machines are 'endowed with':

> What we aim at with machines is to produce an effect that does not require our presence or help, an effect which unfolds without lament and complaint (*klaglos*). What we exclusively aim at are machines, the functions of which, make us superfluous, turn us off, and liquidate us. It is irrelevant that this target condition is only ever approximated. What counts is the tendency, and this is 'without us'.[25]

The atom bomb, 'the brain-child', or rather, *the bastard child* unwittingly conceived by thousands of unsuspecting parents, concentrates power by 'switching off' the critical and emotional faculties of its producers. Having been created 'without' the workers, it can be deployed 'without' their consent. Potential human resistance – the 'consciences' of the collaborators and their impulsive 'freedom' to resist the act or be traumatised by it – are rooted out at source, because the complexity of the nuclear device generates a level of abstraction and detachment that the limited reach of human feeling and empathy cannot bridge. Anders here begins to draw an infernal parallel between two forms of human liquidation that could not be more diametrically opposed: the terrible hell of Hiroshima has its mirror in the terribly comfortable suburbs to which the employees retreat after work. Excluded, that is, 'liquidated' from their own actions these anaesthetised workers cultivate and exercise their own, obsolete private consciences in the hours after work; not least by switching on their television sets (a small anachronism is permitted) and watching from afar how in a distant part of the world a diligently unscrupulous device conducts its work 'without lament and complaint'.[26]

The feeling of having nothing to do with the actual explosion that the screen serves up as a pre-digested 'liquid mush' that streams into the living room is shared by everyone.[27] In fact, this feeling of non-involvement despite participation is the *very condition* for the effect to unfold the way it does, because the reality displayed is socially acceptable on account of a *shared feeling* of personal innocence and non-involvement. As such, a nuclear explosion is the most extreme example of the personally unfelt effects of our habits of consumption and work that routinely contribute to human misery and ecological destruction elsewhere, and that shape the future of our planet. It is precisely this feeling of normality, the 'Blindness to the possible Apocalypse' that turns perception into a 'false witness'.

The real potential nuclear arsenals have to create a posthuman world, a 'world without us' on which there is no life left, is thus the ultimate image of the posthuman world we are already part of and actively helping to further establish. For the bomb powerfully illuminates the removal of the human that underpins the 'oligarchic principle of technology' and the automatic concentration of power it describes. It is here that the Xerox slogan '*in order to predict the future you need to invent it*' reveals its true meaning. As long as these weapons and the knowledge of how to produce them exist, this lifeless world already exists in its germinal state. A desert planet can rocket forth from deep seas or erupt from hidden nuclear arsenals. These weapons have created a 'conditional world', one that has 'not yet' been destroyed.[28] Those who 'control' these objects, however, can undress the future from its indeterminacy, for the world is at their mercy. The future of the world is jeopardised to secure future influence and present interests, and this can be achieved by maintaining a grip on objects that concentrate power all by themselves, without the need for human involvement or consent. This means, as Anders elaborates, that

> the conviction that our world of products is oriented toward the future – a conviction, which stems from the complacent religion-of-progress (*Fortschrittsreligion*) – is completely one-sided. For we defuturise as much future as possible, we drag as much of it as possible into the now to incorporate it into the present.[29]

Referring to the frozen and tinned foods, which allow meals eaten someday tomorrow to be monetised today, Anders stipulates that the power structures that underpin complex technological infrastructures transform the future into 'canned time'. The future 'is taken care of', as Anders sardonically mobilises the slogan of a pension scheme to sum up how existence is attaining an increasingly conditional status in which one must occupy the space of a 'not yet' that is generated by the mere existence of machines that are ultimately serving corporate interests and state power, a process that is unfolding ever more undemocratically without our involvement.[30] What 'matters here', to reiterate Anders's point, 'is the tendency "without us." That this target condition can only ever be approximated is irrelevant.' '*To predict the future*,' one might rewrite the Xerox slogan, '*you need to destroy it*.'

To look ahead to the final section of this book, and the manner in which digital technologies have the capacity to further amplify the 'world without us' Anders begins to think, we can visit the 'Beach at the Edge of Moore's Law', in which Jaron Lanier shows us perhaps the ultimate image of a world in which we have been fully 'liquidated' (despite sitting in perfect comfort): 'Picture this: It's sometime later in the twenty-first century, and you're at the beach. A neuro-interfaced seagull perches and seems to speak, telling you that you might want to know that nanobots are repairing your heart valve at the moment (who knew you had a looming heart problem?).' As Lanier's tale unfolds, we learn that the sponsor of this medical service is a casino, and that the power of the wind is transformed into electricity by 'swarms of leaves [that] turn out to be subtle bioengineered robots that harness that very wind to propel themselves into an emergent shelter that surrounds you. Your wants and needs are automatically analyzed and a robotic masseuse forms out of the sand and delivers shiatsu as you contemplate the wind's whispers from your pop-up cocoon.'[31] All this comfort, however, conceals the radically conditional existence that has been established, for unfortunately, to be part of this world one needs sufficient credit, one must be permitted access to a world belonging to and controlled by others. Back at the beach this might mean, to return to Lanier's image, that when asking for water 'the seagull responds, "You are not rated as enough of a commercial prospect for any of our sponsors to pay for freshwater for you." You say, "But I have a penny." "Water costs two pennies." "There's an ocean three feet away. Just desalinate some water!" "Desalinisation is licensed to water carriers. You need to subscribe. However, you can enjoy free access to any movie ever made, or pornography, or a simulation of a deceased family member for you to interact with as you die from dehydration. Your social networks will be automatically updated with the news of your death."'[32]

NOTES

1. See Anders, 'Die Welt als Phantom und Matrize', *die Antiquiertheit des Menschen 1*, 96–211, esp. 101–105. A drastically abridged version of this essay is available in English translation: 'The World as Phantom and Matrix', trans. Norbert Guterman, *Dissent* 3.1 (Winter 1956), 14–24.
2. See Anders, 'Phantom und Matritze', 103f.

3. William Gibson, 'Google's Earth', *New York Times* (NY edition), 1st of September, 2010, A23.

4. See Jaron Lanier, *Who Owns the Future?* (London: Penguin, 2013) and *You are Not a Gadget* (London: Penguin, 2007). For an introductory discussion of 'nanopayments' see p. 15 of *Who Owns the Future?*

5. Günther Anders, 'Reflections on the H Bomb', trans. Norbert Guterman, *Dissent* 3.2 (Spring 1956), 154.

6. Anders, *atomare Drohung* (München: C.H. Beck, 2003), 146–7.

7. Anders, 'On Promethean Shame', (ch1) 55*.

8. Anders, *atomare Drohung*, 194.

9. This fragmentation of the social body (and individual) is echoed by Deleuze whilst discussing the shift from factory work to what I refer to as 'employment': 'The factory constituted individuals as a single body to the double advantage of the boss who surveyed each element within the mass and the unions who mobilised a mass resistance; but the corporation constantly presents the brashest rivalry as a healthy form of emulation, an excellent motivational force that opposes individuals against one another and runs through each, dividing each within' ('Societies of Control', 179).

10. It is worth giving this passage in full, as it succinctly reiterates the 'Promethean conception' of technology: 'Nothing is more misleading than the assertion that machines are "*morally neutral.*" This particular "Philosophy of Technology" is rarely explicitly elaborated, but it is taken to be self-evident. It assumes that machines are at our disposal, that it merely matters *how* we use them. The only thing that is of consequence what purpose we give them: a moral or immoral, humane or inhumane, democratic or antidemocratic one. This popular idea needs to be combatted. [...] After all, we are always already being shaped by the very existence of a particular machine. For, each machine either prescribes a particular relation between us and our fellow human beings, between us and all matters, between all matters and us, or the machine "posits" and "sets the ground" for this relationship to unfold. Each machine shapes us in this manner regardless what particular purpose we deem to be using it for or what we plan to achieve with it. Yes and it does so, regardless from within which political and economic system we turn to it. This means: every machine is in itself already its use' (Anders, *Antiquiertheit 2*, 216).

11. See Bernard Stiegler, *The Decadence of Industrial Democracies: Disbelief and Discredit, Volume 1*, trans. Daniel Ross and Suzanne Arnold (Cambridge: Polity, 2011), 9ff.

12. 'Nobody can be accused', as Anders puts it in *The Obsolescence of Human Beings 2*, 'that they have the aim to make humans unemployed, not even an entrepreneur who is fanatically reducing his workforce. What entrepreneurs want is *not unemployed workers, but a business that does not require employees.*' The effect of this 'trend' is that the world becomes 'overmanned' the more technology improves. See 'Einleitung: die drei industriellen Revolutionen' (Introduction: the three industrial revolutions), *die Antiquiertheit 2*, 15–33, 26–27.

13. 'The Great Decoupling' describes how since the 1980s technological change has been a driver of steadily rising productivity, GPD and company profits whilst simultaneously leading to lower employment figures and sinking mean income. This

gulf, as Brynjolfsson and McAfee predict, will further accelerate as computing power and ability continue to develop exponentially in the future. See Erik Brynjolfsson and Andrew McAfee, 'The Great Decoupling', *International Herald Tribune*, 12 December 2012, 8. See also Brynjolfsson and McAfee, *The Second Machine Age*, esp. 98–185. The oligarchic principle is echoed by Lanier when he suggests that those closest to the fastest computers with the most access to everyone's information hold the most power. See, for instance, *Who Owns the Future?*, 27.

14. See Anders, 'On Promethean Shame', (ch1) 82*.

15. Ibid., 90*.

16. Harry Truman, 'Statement by the President of the United States (6th of August 1945)', accessed via http://www.trumanlibrary.org/whistlestop/study_collections/bomb/large/documents/index.php.

17. Anders initiated an exchange of letters with Claude Eatherly, a weather reconnaissance pilot who was wracked by feelings of guilt after being unwittingly involved in the deployment of the atomic bomb on account of transmitting his weather report of the target area. In the opening letter, Anders presents Eatherly's case as being symptomatic of a shared, but unacknowledged condition of 'guiltless guilt': 'The "technification of our being": the fact that today it is possible that unknowingly and indirectly, like screws in a machine, we can be used in actions, the effects of which are beyond the horizon of our eyes and imagination, and of which, could we imagine them, we could not approve – this fact has changed the very foundations of our moral existence. Thus, we can become "guiltlessly guilty," a condition which had not existed in the technically less advanced times of our fathers' (p. 1). See Anders, *Burning Conscience: The Case of the Hiroshima Pilot, Claude Eatherly, Told in his Letters to Gunther Anders* (New York: Monthly Review Press, 1962). The book and Anders's presentation and 'exploitation' of Eatherly, who was at the time incarcerated in a military hospital after being declared insane by a military tribunal, was highly controversial and was heavily criticised in the media, see for example Ronald Bryden, 'The Myth of Major Eatherly', *The Spectator*, 16th of February, 1962, 5. Anders's own representation of the hostile responses and the resulting travel ban to the USA forms the 1982 introduction to his *Hiroschima ist überall.*

18. See Joseph Masco, *Nuclear Borderlands* (Princeton: Princeton University Press, 2006), 267ff.: 'Since 1943 a whole industrial infrastructure – a nearly $6 trillion national project and one of the largest industrial enterprises in U.S. history – has been shielded from public discourse. Consequently, nuclear secrecy evolved during the Cold War into an enormous state apparatus. […] At times,' as Masco elaborates on how secrecy is used to conceal health and environmental effects of nuclear research and fend off potential lawsuits, '"national security" has meant protecting the U.S. nuclear complex from U.S. citizens, rather than from foreign adversaries.'

19. Anders, 'H Bomb', 150.

20. Ibid., 149.

21. Anders, *die Antiquiertheit 1*, 247.

22. Anders, 'H Bomb', 150.

23. Anders, *atomare Drohung*, 200.

24. 'There is no situation that is morally more evil', Anders explains what I call 'anaesthetic violence', 'than the situation in which evil has been so fully integrated

into the situation that the individual itself is spared from being evil. [...] The machines that populate our world have spared us from incurring guilt, just as they have relieved us from baking bread or from calculating statistics' (*atomare Drohung*, 88). On this point also see my 'Desert Ethics', esp. 42–4, 53.

25. Anders, *atomare Drohung*, 199.

26. Hans-Martin Lohmann succinctly introduces the trajectory of Anders's perspective on our private conscience in a response to the First Gulf War: 'One is opposed to acts of violence that become openly manifest because they are atavistically uncivilised ("No blood for oil"). In doing so, however, one stubbornly ignores that in each case the individualistic practices that mark ordinary capitalism – starting with the thoughtless production of waste to the ruinous competition to acquire the newest status symbols – add up to a chain of violence (*Gewaltzusammenhang*) that, admittedly, has "invisible" long distance effects' (p. 218, my translation). See Lohmann, '"Glückliche Zeiten als es noch Kriege gab." Variationen zu einem Satz von Günther Anders,' *Günther Anders Kontrovers*, 211–20.

27. Anders's repeated insistence that information technologies provide us with 'liquefied content' that is easy to consume underlines that this liquidation and the processes by which our critical faculties are 'switched off' also describe our relationship to media technologies.

28. Anders, *atomare Drohung*, 123.

29. Ibid., 124.

30. Ibid., 123.

31. Lanier, *Who Owns the Future?* 8–9.

32. Ibid., 14.

Chapter 4

Invisible Monsters

Your Smartphone is an Atom Bomb

> Without knights no chivalry, without court no courtliness, without a parlour no charm, without material support no consideration and deference can last indefinitely, not even as pretence. Just like the subtlety of the private lives of our souls shrivels up in a world that cheats us out of leisurely idleness and the other preconditions of privacy.[1]
>
> **Günther Anders**

> The invisibility of the monsters is the monstrosity of today.[2]
>
> **Günther Anders**

In July 2015, The Future of Life Institute issued a widely publicised open letter on 'Autonomous Weapons' and the potential of artificial intelligence and robotics research to lead to a global arms race comparable to that of the Cold War. 'The stakes', as the letter reminds us, 'are high'. Within 'years, not decades', autonomous weapons systems that can 'search for and eliminate people meeting certain pre-defined criteria' will be 'practically if not legally feasible'. Such weapons will pose an additional threat, the letter goes on to explain, because unlike nuclear weapons, they are likely to be highly democratic insofar as their creation will be comparatively cheap and not contingent on access to rare and highly controlled raw materials, secret technological knowledge and scientific research. If developed, 'it will only be a matter of time until [such weapons] appear on the black market and in the hands of terrorists, dictators wishing to better control their populace' or until they are used for 'subduing populations and selectively killing a particular ethnic group'. It is hard not to sign up to the letter's argument that 'starting a military arms race' leading to weapons 'beyond meaningful human control' is 'a bad idea' and that it should be prevented from happening.[3] In view of my above discussion of Anders's notion of human liquidation, however, the letter's transfixion on artificially intelligent *weapons* (that are capable of autonomously liquidating humans) might appear to be far too narrow.[4] From the perspective Anders opens up, digital technology is not dangerous because it might lead to more autonomous and potentially even fully emancipated artificially intelligent weapon systems. A smartphone is like an atomic bomb not because the technological innovations that make it work might be weaponised, but because it mobilises and even amplifies the inherent tendencies of complex and autonomous technology: it facilitates 'the escape of inhibition and scruple' that Anders begins to isolate. The danger, therefore, is the further removal of the human. Even though one is not fully 'present to', or at times even radically absent from, one's activities as an employee, at least one knows *that* one is working. Today, as the big data revolution is in full swing and our digital devices innocuously and indiscriminately collect data that can be mined for future applications, perhaps even in weapons systems, this is no longer the case.

The Second Machine Age by Brynjolfsson and McAfee (who are both professors of digital business at MIT) is my point of departure for the somewhat polemic view of the digital age I present here. Brynjolfsson and McAfee suggest that we have passed 'the inflection point' beyond which even sophisticated 'human' activities such as writing poems and newspaper articles, driving cars, policing cities or offering a medical diagnosis can be undertaken by computer algorithms.[5] The mere existence of these abilities perfectly illustrates just how limited our foresight is with regard to the capabilities and effects of future technologies. On their own admission, the very same authors, as well as a number of experts previously came to the conclusion that such work could perhaps never be conducted by machines. In a 2004 study, for example, David Autor, Frank Levy and Richard Murnane noted that non-routine work such as 'truck driving, medical diagnosis and persuading and selling' will remain beyond the capabilities of a computerised task (all of which are now realities).[6] Even at the outset of the twenty-first century, it was apparently still hard to imagine that software such as the Amazon recommendation algorithm would become a more persuasive bookseller than any human being could ever be. The machine, however, did not suddenly attain human qualities and become intuitive, intelligent and sensitive enough to replace the antiquated human bookseller. It merely began to access data sets that are generated by our own behaviour while buying books online. This data feeds (self-improving) algorithms that calculate what titles are statistically the most likely to initiate a purchase. The potential buyer, of course, experiences this mathematical operation quite differently, namely, as a captivating display of books that seem too exciting to be left on the virtual shelf. Even if these books are not bought, the interaction with the webpage will further quantify and categorise the inventory and make an infinitesimally minute contribution to its monetisation.

This basic example can serve to illustrate what makes Anders's idiosyncratic understanding of human obsolescence so perceptive, and this arguably even more so today than when it was conceived. Paradoxically, more conventional, or what one might call 'more rigorous, non-exaggerated' approaches that measure and rely on the 'fantastical' realm of our perception that Anders deems obsolete, portray a tendency to think machines *too humanly*. Autor, Levy and Murnane were tempted to assure us that a computer could never be as persuasive as a human seller, because it seemed improbable that it could acquire the interpersonal skills and spontaneity of a human being. It was assumed that a machine would only be able to replace a human worker if it acquired humanlike qualities. In 2004, the mechanised bookseller was still imagined to be a kind of sales robot that would welcome us in a book shop with a friendly 'hello', break the ice with a joke, and then help us choose a book.

It is here that the conceptual framework that underpins Anders's diagnosis that machines harbour a 'world without us' because they 'create an *effect* without human involvement' isolates what the divergent approaches resting on comparisons between machines and humans seemingly cannot. No, we are not becoming obsolete because machines are becoming 'smarter' and ever more capable (although this is also a consequence of the tendency Anders isolates). We are becoming obsolete because we are increasingly expressing ourselves with the aid of digital devices, because we are willing to make compromises and live in accordance with a machine's limitations and demands. The so-called ELIZA effect, which names our 'desire to cover for a robot [or another machine] in order to make it seem more competent than it actually is', is perhaps one of the most immediate markers of the 'hubristic self-humiliation' and unprecedented 'selflessness' towards machines that Anders describes in 'On Promethean Shame'. For in order to make the machine work, we curtail our own range of expression so as to ensure that the machine will respond and perform in a manner that makes our lives easier.[7] Put otherwise, and in the words of Anders, we seem willing to exert effort so as 'to not function as ourselves', turn away from humans and adapt our behaviour to allow the machine to serve us to the best of its ability.[8] By doing so, we tie ourselves ever further into infrastructures, adapt our habits to technological solutions and expose ourselves to increasingly undemocratic constellations of power and big data applications that are actualising a version of the future in which human contributions are potentially no longer required. As we act out, enjoy and experience our spontaneity and creativity with the help of digital technologies, we also generate the data that machines require to analyse and mimic our behaviour, and to steer or modify it. Far from being malignant, this adjustment to our behaviour is enforced by the anaesthetic tendency of the machine itself. By making our life 'easier', more comfortable and convenient, more exciting and organised, the device renders the very idea of living without it nonsensical. Anders's perspective, of course, does not discount the fact that technological progress brings great benefits and opportunities, also for the flourishing of individuals. 'What counts' for Anders, however, as I have noted several times already, 'are not reasons, but, rather, what is actually done: in order to combat the limitation of their liberty, humans limit themselves in a new way'.[9] His view is certainly highly unbalanced, but it is so *precisely because* the effects of machines that radically surpass human capabilities cannot be counter-balanced by perfectly benign human intentions.

The book recommendation algorithm, to return to my example, is not a better bookseller because of its superhuman, artificial intelligence, or because it manipulatively pressures us into doing something we do not actually want to do. Its persuasive power, rather, is based on our own production of data. Put otherwise, it is a product of the 'work' of enthusiastically browsing books online, of searching for, selecting and ignoring products on display. With the help of an unlimited data set, the algorithm perceives what we, as limited creatures, cannot. The bookseller is hence in danger of becoming obsolete because his salaried employment, his knowledge, spontaneity, experience and intuition are no longer required to sell books. We provide these emotional energies ourselves while spending time online. Because we supply the 'human experience' that machines need to feed information back to us in a processed form, ever more of these human qualities are in danger of 'escaping us', to use Stiegler's term. We are of course rewarded for our unsalaried effort by the interface itself, as it provides us with a shopping experience that is entertaining, more affordable, and not restricted by the limited shelf space of physical shops. The work of selling books, however, is *not taken over by the algorithm, this work is simply abolished* because the existence of this algorithm effects a large scale shift in consumer behaviour.

In response to Evgeny Morozov's discussion of personalised content and e-readers we can take this image further.[10] Today, as our reading patterns, preferences and other data concerning our reading habits are recorded while we interact with e-books, the data and meta-data required to algorithmically modify the vocabulary of a text to fit personal preferences is also gathered. As computer-generated poems are already winning poetry competitions, it seems likely that highly readable literary objects with the statistically highest probability of fitting the personal tastes of the reader 'will become available' at the proverbial 'push of a button' in the not too distant future. To state that such a text was actually '*written*' would be too human an expression. As smart as our smart devices may appear to be, they do not require 'artificial intelligence' or humanlike intuition and emotion in order to mimic human behaviour and exhibit signs of human creativity. They merely need the processing speed to be able to access and compute the exponentially growing patterns of behaviour documented by data sets that are stored while we interact with networked devices. An Andersian exaggeration might conclude, therefore, that today not only has the distinction between the bookseller and buyer become obsolete – the buyer recommends and sells him or herself books while interacting with the seller's website – the distinction between author and reader is also beginning to look antiquated. Roland Barthes's famous diagnosis that 'the birth of the reader must be at the cost of the death of the Author' is perhaps on the verge of turning into a simple description of the effects of a technological object that can autonomously generate literary content.[11] This would finally eradicate the ambiguity with which Barthes's essay has challenged students of literature for decades.

This opening illustration has hopefully helped clarify how the *unpaid employment* that Anders invites us to think is neither simply a form of 'shadow work', to use Ivan Illich's famous term, nor a variant of the 'free labour' that Tiziana Terranova has described. It is not directly comparable to the unpaid efforts of stay-at-home mothers that happen 'in the shadow' of the economy without ever being statistically recorded or socially recognised as work, to use Illich's example.[12] Neither is it work that a consumer or service user does for 'free' in the place of an employee, which is the case when we check into our flights online, use an automated check-out or create online content by writing blogposts, uploading pictures and videos or spend time on social media.[13] Anders's notion of 'unpaid work', to use one of his formulas, 'is much more and much less' than these forms of free labour.[14] It is 'less' because we conduct this work merely by interacting with machines. It is 'more' because the accumulating data sets have the capacity to lead to machines that can definitively move human activities and capacities out of the reach of human beings. To get to the heart of unsalaried work, and the close connection that exists between this activity and Promethean shame, I end on a discussion of speech recognition, the machine capability that perhaps most overtly encapsulates the tensions that Anders's thought hinges on.

In 2011 the American game show *Jeopardy* was won by Watson, a computer contestant developed by IBM that used advanced speech recognition software. Watson was entered against the all-time *Jeopardy* champions Ken Jennings and Brad Rutter, who had both won the game many times before, earning millions of dollars in the process. This time, however, the machine won, answering more questions correctly than the two human contestants combined.[15] Because *Jeopardy* does not seem to follow fixed rules like chess, but revolves around general knowledge questions that include puns, rhymes and wordplay, because the contestant needs 'to get the joke' to even understand the question, competing in *Jeopardy* appears to be one of the non-routine tasks that in 2004 would have been declared impossible for a machine to complete. Before the advent of big data one did not fully realise that machines do not need to acquire humanlike qualities in order to potentially outperform the human in its own supposedly exclusive domains, such as speech, humour and general intellect. The software can do so because of vast data sets that can be statistically evaluated. In the case of Watson, this data does not merely comprise of digitised content in the form of scanned books, *Wikipedia* and other primary sources, but also the metadata that records how humans interact with this content and in what contexts specific patterns of speech are deployed. Watson does not 'get the joke'. The programme can merely process enough data to calculate that it is statistically likely that the question contains the elements of what we humans call a joke or pun. This, in turn, is possible, because programmers know that wordplay, verbal wit and humour are important components of the game. One can 'count on' the need for such spontaneity in advance and begin to quantify and predict it. Once such co-ordinates are available and the programme can be trained and improved by analysing thousands of hours of recorded game play, the machine begins to seem as if it is performing spontaneously in a superhuman humanlike manner. Put otherwise, if sufficient data sets and computing power are available, then supposedly non-routine tasks transform into routine ones.

The Second Machine Age gives us a good idea of what provided the raw material, that is, the data and content that animated the machine-brain: 'unpaid work'. As the two MIT professors for digital business who authored this book celebrate the trends Watson encapsulates, the following passage should be read with an inner voice that is evangelically cheerful, and not with the cynicism that inevitably reverberates through these lines if viewed through the lens of Anders's writings:

> The old business saying is that 'time is money', but what's amazing about the modern internet is how many people are willing to devote their time to producing online content without seeking any money in return. [...] All this user-generated content isn't just making us feel good by letting us express ourselves and communicate with one another; it's also contributing to some of the recent science-fiction-into-reality technologies we've seen. Siri [Apple's speech recognition software], for example, improves itself over time by analysing the ever-larger collection of sound files its users generate when interacting with the voice recognition system. [...] For a while [Watson's] database also included the salty language-filled Urban Dictionary, but this archive of user-generated content was removed after, to the dismay of its creators, Watson started to include curse words in its responses.[16]

It is not hard to envisage what images Anders might have arrived at in response to machines that unwittingly insult their users, without knowing how it might feel to be insulted. The example of speech recognition and Brynjolfsson and McAfee's pragmatic observation that users do not seem to expect a financial reward for an activity that contributes to the improvement of an application, leads us to the heart of the phenomenon of unsalaried work.

In 'The Obsolescence of Privacy', a 1958 essay that builds on 'The World as Phantom and Matrix', Anders suggests that each device that 'delivers and serves up' (*beliefern*) information for our consumption also turns into an '*Auslieferungsgerät*' (p. 210), that is, it becomes 'a device through which we expose, deliver and surrender ourselves' to the phantom worlds animated by machines.[17] More importantly, in doing so we also surrender ourselves to those who profit from, run, and thus ultimately own these machines. Today, as we willingly share our 'private' lives online and smartphones disregard our privacy by default by discretely tracking and recording the way we interact with them (at times, to interact with other human beings), Anders's essay on the effects of McCarthyism, state surveillance and commercial profiling seems 'prophetic'.[18] It describes the actual activity that 'unpaid work' is comprised of. Machines of self-surrender give us '*the homework of becoming a collaborator* (Mitarbeiter) *in our own de-privatisation by relinquishing our sense of shame and embarrassment*' (p. 237). Our work consists of self-exposure through the voluntary, consensual (but often unregistered and subliminal) activity of providing information about ourselves, even if this information is only ever processed by algorithms and never meets human eyes. This activity not only provides the data through which 'knowledge escapes the human' by contributing to applications that mimic human traits, this indiscriminate 'surveillance' also cultivates new vulnerabilities, not least in the form of ever more personalised and captivating online services, the use of which increasingly structures our lives. 'It is all too easy to forget that "free"', as Lanier reflects on the epithet of so many applications, 'inevitably means that someone else will be deciding how you live'.[19] It is this voluntary but tacit surrender of self-determination that Anders seeks to capture in his discussion of his particular account of unpaid work.

Anders opens his reflection on the work we conduct to 'de-privatise' our existence by gradually reconfiguring and combatting the feelings of embarrassment, modesty and discretion that mediate our private sphere, with an example that will seem entirely harmless. He discusses television shows that access homes 'through the key hole' to investigate, question and openly display the interior, intimate lives of their inhabitants. For Anders, such shows openly portray that in a mediated world the motto '*each to their own*' (Jedem das Seine) is replaced by the catchphrase '*to everyone what is mine*', or even '*everyone is here for me*'' (Jedem das Meine, jedermann für mich). This often proud act of displaying oneself, conceals that this exposure is an act of expropriation through which one's existence is turned into *Allgemeinbesitz*, into 'common property'. Anders describes the social pressure to turn oneself into the kind of person who has something to show – photographs, artistic or academic achievements, a tidy home – '*a self-expression racket*' by means of which machines, by offering us a stage, gently pressure us to join the 'fraternity of "the creative ones"' (pp. 234–235). In view of social media, Anders's exaggerated images have turned into simple observations. Yet, this self-exhibition is an inconsequential by-product of the true act of exposure. For the homework of relinquishing our sense of shame and embarrassment at being immodest and showing off, ultimately consists of acquiring and cultivating '*the malaise of being unique*' that Anders addresses in 'On Promethean Shame'. In order to compulsively share everything we feel, do, and think, we need to *feel* – on some level at least – that the condition of singularity that marks human existence has no value. We must see in it an abnormal, embarrassing trait that must be combatted. By de-privatising ourselves, however, we begin to lose the protective boundary we acquire by default in the process of articulating ourselves with the help of artificial media; we lose the sense of discretion which, despite often working to our disadvantage, nevertheless acts as our ultimate retreat.

As Promethean beings, the boundary between our internal and external lives is permanently in flux, and cultivated by our relationship to shared media that we have limited agency to shape (language and fashion are perhaps the most intuitive examples). Our agency is limited, precisely because we are not, and cannot be, the sole owners of the medium by means of which a self is cultivated through articulation. We can learn to speak English, and attempt to make sense of our experiences by bringing them to expression, but we cannot own the medium that enables us to do so. Put otherwise, we can never speak this language well or perceptively enough to fully eradicate misunderstanding, fully present ourselves and fully satisfy our desire to speak. Far from being a failing of the medium, the limited grasp we have of language – the tendency language has to resist our attempt to master it or to show us up as inadequate speakers – also carves out the protective hollow we call 'the self' (for want of a better word), this unspeakable, inaccessible and unspoken experience of singularity that reverberates in every expression.

Language is a mark of expropriation – by learning it we are inscribed, and we inscribe ourselves into, national and local idioms. We expose ourselves to state power, social codes and rules of politeness and thus tacitly acquire a set of values, affective dispositions and affinities. We become 'English' in more than just a linguistic sense, because the language itself will play a part in how our experiences and feelings are registered and expressed. This expropriation unfolds, however, by also generating a retreat, what Anders calls a '*Binnenraum*', an *interior space of the self* that cannot be owned (not even by the self) as it is 'landlocked' by the intangible medium itself (p. 219). Regardless of how hard they might try, state power and economic corporations cannot 'own language', they can merely influence the way we use it and what values, desires, affects and experiences we associate with the words we use and hear. Put in terms of Louis Althusser's discussion of ideology, power and influence need to be exerted through acts of 'interpellation'. Subjects are 'recruited' and invited into specific points of view, roles and subject positions through an act of recognition that makes them cry out '(aloud or in the "still, small voice of conscience"): "That's obvious! That's right! That's true!"'[20] Our acquisition of intuitive responses and the sense of ordinariness that normalises social hierarchies and world views are partly guided by institutions that teach us to speak 'the right way' to set the ground for such acts of recognition to unfold.[21] Interpellation, in effect, combats singularity with aesthetics. It aims at immediate and intuitive identifications with larger social groups or subject positions, and inscribes social relations into the very transparency of language itself.

In light of Anders's discussions of *Weltfremdheit*, we can add that interpellation can only happen because of our non-specificity, because as indeterminate and indefinite beings we are malleable, receptive for interpellation because we need to make ourselves into something. This means, however, that we also resist interpellation by default of a singularity that cannot be fully eradicated. Put otherwise, the fact that we must be recruited by an act of recognition also means that we at times do not recognise, or do so too late. The impulse to deem something 'obvious' does not reliably set in. For Anders, we resist power by default of our own singularity, we will and can never become fully compliant with the positions we are encouraged to adopt. Not only might we be too imperceptive, distracted or inarticulate to recognise unspoken expectations, but our own experiences will shape the way we interpret (and resist) certain social roles and conventions through misunderstanding and on account of impressions that experience might suddenly begin to question. Interpellation might shape our subjectivity, at the same time however, this external pressure animates a counterforce, the retreat of the self into its own 'landlocked' interior, the private self, sheltering behind its external façade. This space, as Anders suggests, cannot be accessed by state or corporate power as the surface protecting it cannot be pierced; the self is only ever evaluated from the outside (even by the self). As such, the very media through which we are encouraged to perform certain subject positions and replicate world views also become protective barriers. They preserve the possibility that we might not mean or believe what we say or communicate, or may no longer do so in the future. We might also develop ways of speaking, thinking or dressing that resist and challenge the obviousness and transparency of the facts about the world we intuitively acknowledge. Moreover, we might share and cultivate these private views with select others, but not with those whose position might be questioned, ridiculed and subverted by them.

This process of singularisation takes on a radically different dynamic the moment an intangible or passive medium that cannot be owned – language – is replaced with an active registering machine, what Anders refers to as a '*abhör-Apparat*', literally a 'bugging device' (p. 216). For such devices, by definition, are always owned by someone else. They do not belong to the individual who is their apparent proprietor, because these machines also pass on information to another entity. Put otherwise, language spoken through a bugged phone is no longer simply speech. Anders uses a discussion of the surveillance techniques adopted in McCarthy's America, and the normalisation of being under surveillance as a means to develop his notion that information technologies seek to disarm and invalidate the 'landlocked' interior self. For not only do bugging devices attempt to catch out the 'private' self behind the mask by intruding into the private sphere of individuals, they also create a state of affairs in which the insistence on privacy becomes suspicious in itself. It is here that the voluntary self-surrender through self-exhibition that combats the impulse to be private seamlessly feeds into an act of dispossession. The moment embarrassment and a desire for privacy and discretion are perceived to be shameful individual flaws that hold us back, the moment they are deemed to possess no value and the more we combat them as such, the more we begin to institute a situation in which these very qualities become suspicious, appearing as a pretext to hide private transgressions, vices, or even illegal activities. '"I have nothing to hide" is proof', as Anders explains in response to the now ubiquitous slogan that began to appear in McCarthy's America, 'that the expression of *shame* (i.e., "to be in need of shame") *is being identified with immorality,* and *shamelessness* ("to no longer be in need of that feeling") *with being moral*' (p. 229). What this new moral codex conceals, is that it is not only privacy that is relinquished or at least radically transformed in its expression, but the very nature of private property itself. For without realising it, *the de-privatised individual that has outsourced itself, recorded itself, quantified itself and thus given itself to the machine has also become an expropriated individual.*

This act of expropriation – Anders provocatively redirects McCarthy's anti-communist fervour – establishes a new form of common property, and hence a form of socialism and state-ownership. This form of socialism fully protects what we usually call our '*private property*' (Eigentum), our cars, phones and the 'things' we have accumulated. To show us what we are asked to give up, Anders cites a newspaper report that contains the statement of a private investigator: '"We only tap phones of wives [...] if we are certain that the actual telephone belongs to the husband."' According to this logic, Anders concludes, '*every word that is spoken through the "telephone" belongs to the proprietor of the object, that is, it belongs in this case to the husband*' (p. 231). If the wife were to insist on her privacy, not only would she automatically incriminate herself by implying that she has something to hide, she would also be claiming ownership of something that is not hers to own: *the very words she has lost ownership of because they were spoken into a receiver belonging to her husband, and now ultimately reside with whoever owns the recording*. This seemingly absurd image is a simple description of the way in which the data we generate is treated today, for we share the position of the wife in the example, although the phone is not owned by some husband, but by service providers, software companies and the state. At best we can be said to 'own' the actual hardware we use (even though our gadget would be useless without the services it allows us to access). We do not, however, exclusively own anything that passes through these machines. Rather than being proprietors of ourselves, and of everything that emanates from and marks our own singularity, we have unwittingly given ourselves freely to software companies, service providers and, of course, to state and corporate supervision, which can use all this information to profile, quantify and calculate our singularity. Our singularity no longer retreats behind a public façade, but is revealed, classified and is analysed in ways that we cannot. Although machines cannot know the self any more than we do, they can, as the slogan goes, suggest that they know us better than we know ourselves, for they reveal patterns that remain invisible to us.[22] The moment we begin to express ourselves via machines that record our behaviour or wear them on our wrists to track our health, our words no longer evaporate into thin air or the porous, imperfect memories of our interlocutors, our pulse and bodily activity is no longer unmeasured and inaccessible, discreetly concealed under our skin. With this data, machines can circumvent their lack of access by revealing what Anders calls the 'soft parts', the parts of us that are still malleable, elastic and receptive to being shaped by others.

The human engineer, as Anders reminds us, does not investigate the body to know what it *is*, but rather to find out what it can still become. In analogy to this we might say that algorithms that process data do not only seek to classify what we are, but also what we might still be in the future. Our 'soft parts' might reveal ways in which we can become more receptive to certain services or products than we are now; our health data might show that we should be prodded to do exercise so as to reduce the risk of becoming a burden for our health insurance provider; our online behaviour (or avoidance of online services) might suggest that we are to be placed on a watch list, because we are involved in criminal activities or are preparing future acts of violence, etc. In contrast to acts of interpellation that seek to influence us in at times obvious ways – advertising does not conceal its intentions – machines of self-surrender threaten to be persuasive in ways that we cannot possibly comprehend and counteract, for they exploit our limited self-knowledge. Not only might they predict who we will or can be, and prod our 'soft and vulnerable parts' to ensure that this prophecy becomes true (or, in the case of suspected criminality, that it does not happen), they also block certain ways of being in advance, limiting what we can only learn by experience, that is, by making mistakes, being imperfect and colliding with (but also overcoming) our own limitations. The moment this form of potentially painful self-knowledge is devalued, the moment we trust the display more than we trust ourselves, we have definitively declared that 'our own business is no longer our business', we have placed ourselves into the hands of applications and faceless authorities that we trust at our peril.[23] As we freely provide data and teach machines how to speak, it is not just speech that threatens to escape us and be passed over to machines more eloquent and witty than we are; the very foundations of self-determination and self-knowledge that human expression and democratic societies rely on. For when we believe that machines know us better than we know us ourselves, there is not much left for us to do, think, feel or decide. What counts here, of course, to use Anders's reminder, is again the 'tendency' towards expropriation, and not whether the endpoint of a totally expropriated individual could ever be achieved.

Anders sums up our dependence on machines by suggesting that 'the more power is integrated, the more silent its command becomes'.[24] It is to this problematic, I believe, that 'On Promethean Shame' offers a crucial perspective. For it reminds us that the feeling of nudity and curtailment that can befall us when we have learnt to rely on a machine is a pain we suffer alone and in silence. The feeling of deprivation that Katherine Hayles sums up by suggesting that a missing phone signal makes her feel as if her 'hand had been amputated', or the concomitant examples Sherry Turkle cites in *Alone Together*, where one of her interviewees describes feeling '"like [she] had lost her mind"' upon misplacing her phone, powerfully illustrate how our relinquishment of privacy is enforced through our feelings of discomfort, inconvenience and exposure.[25] Anders sees in the tendency machines have to blackmail us with feelings of curtailment, solitude and self-contempt, the workings of power in a world in which ideology has become obsolete. For if we lose access to the service they provide, we fall victim to a kind of 'panic' that we seek to avoid at all costs. While reflecting on how this panic is a product of our innate tendency to embody technological objects, Anders evokes the Prometheus myth and presents these addictive, stimulating machines as 'unfelt chains'. Because 'it belongs to the nature of these chains that they are not felt by those who wear them', their hold is revealed only when we lose them. 'If we were put in a "chainless" state', Anders elaborates, 'if all the machines were requisitioned that deliver the world to us in a liquid state as if from a mother's breast [...] then we would suddenly be faced by the world in form and object. In a consistency, therefore, that does not let itself be liquefied for immediate consumption'. Stripped in this manner, 'we would not be able to truly comprehend our absolute dependency on the processed and machined world', but it is likely that the 'kind of panic would begin to well up that we feel in view of starvation'. We suffer 'the panic of a toothless man', as Anders concludes, '*who, accustomed to being fed with mush, sees himself confronted with death from hunger on account of sitting amongst breads, apples and salamis*'.[26]

The typically polemic and moralising image of a toothless man succinctly sums up the antithetical tendencies that shape an anaesthetic life. The more machines make us feel good and empowered, the thinner our skin becomes. The more they enhance our lives and offer us control, the more they have the potential to live our lives in our place and put us under the supervision and control of others. The more we use them to trick the body into agency and out of boredom, the more we 'lose our bite' and blunt our critical awareness. The more freely we surrender ourselves to those who profit from our precarious sense of well-being, the less we realise that through our reliance we give ourselves to the 'owners' of the devices that we believe we possess. In a highly technologised country, as Anders concludes in *Human without World*, 'the "chains" that bind [us] to a world belonging to others have been made so comfortable and smooth, that we take the chains to be the world, and even believe that they are our own world'. We do so to such an extent, Anders elaborates, that it seems clear that 'we can already no longer imagine another one, [...] and so defend this world as ours with tooth and nail'. It is in this act of 'claiming as a holy right' the act of surrendering oneself to a world belonging to others that Anders sees clear 'evidence' that our existential condition is being factually realised. We are becoming 'worldless', and this, if we are fortunate enough to live in comfort, without realising it.[27]

In the course of their optimistic analysis, not even Brynjolfsson and McAfee disregard the fact that our increasing dependence on and interaction with digital technologies has the potential to create unprecedented levels of inequality between the richest and the poorest inhabitants of our 'shared' world. Using *Wikipedia* as an example, however, they focus on opportunities and present our increasingly connected world as a truly collective human achievement. This image fails to capture the fact that this collective enterprise entails an activity that is not recognised as such, for unlike *Wikipedia* contributors, there is no visible end product for our unpaid activity, which means that *we do not know and cannot see what the data we generate will be used for*. It is likely that this data will remain for ever unaccessed on some server, but it might also make an infinitesimally small contribution to the improvement of a speech assistant that costs thousands of jobs or to an application that saves countless lives. It might also simultaneously feed into the development of an operating system for one of the artificially intelligent autonomous weapons that will likely come into existence in the not too distant future. Regardless of the use to which the data is put, the chain of abstraction between the human interaction with a networked object, and all of the deeds that autonomous machines may be able to perform on the back of the data this use generates, is so elusive that it seems probable that it can only be registered by a powerful algorithm.

The exuberant celebrations of a future in which 'you'll see collective intelligence [...] come not only from networked brains but also from massively connected and intelligent machines', to cite one of the endorsements of *The Second Machine Age*, routinely neglect to mention that the 'brains' that networks connect do not represent humanity, and neither does the data that these brains generate.[28] They merely represent those societies affluent and 'developed' enough to use the internet on a regular basis. It is a collective, therefore, only of those who regularly interact with digital devices, of those who have been shaped by this use and have built their lives around such machine-mediation. In view of this problematic that threatens to tacitly set a new 'human' standard by default, because only certain humans contribute to data and thus to the emergence of future machine ability, the financial inequality that the expropriation of data generates seems only one part of the problem. Although this inequality, as Lanier reminds us, is real enough:

> The clamor for online attention only turns into money for a token minority of ordinary people, but there is another new, tiny class of people who always benefit. Those who keep the new ledgers, the giant computing services that model you, spy on you, and predict your actions, turn your life activities into the greatest fortunes in history. Those are concrete fortunes made of money.[29]

If we begin to adopt the perspective that enthusiastic technologists like to ignore, that is, the viewpoint of limited and potentially obsolete humans, then it may be possible to shake the veneer of normality that mediates our commerce with machines that record our every keystroke, eye movement and step. Even if it does not become less ordinary to use digital devices, it may be possible to appreciate – at least in thought – that these machines seem far more inconspicuous than they actually are, and that it is this deceptive sense of ordinariness that induces obsolescence by numbing our critical faculties. Anders's writings suggest that the feeling of normality, which Kracauer suspects to be the product of 'powerful forces [...] anxious to prevent anyone noticing anything here', is ultimately rooted in our Promethean frailty, our plasticity of feeling that transforms the 'artificial world into what seems to be "Nature," the *a posteriori* into the seemingly *a priori* and the conditioned world into the seemingly necessary one'.[30]

To what extent Anders was writing for 'the day after tomorrow', that is, for today, becomes apparent if we listen to the public admission of 'Promethean Shame' provided by the all-time *Jeopardy* champion (category: humans) Ken Jennings, in his description of what he felt while being beaten by a machine:

> [W]atson won handily. And I remember standing there behind the podium as I could hear that little insectoid thumb clicking. It had a robot thumb that was clicking on the buzzer. And you could hear that little tick, tick, tick, tick. And I remember thinking, this is it. I felt obsolete. I felt like a Detroit factory worker of the '80s seeing a robot that could now do his job on the assembly line. I felt like quiz show contestant was now the first job that had become obsolete under this new regime of thinking computers. And it hasn't been the last. […] All I know is how it felt to be the guy put out of work. And it was friggin' demoralizing. It was terrible. Here's the one thing that I was ever good at, and all it took was IBM pouring tens of millions of dollars and its smartest people and thousands of processors working in parallel and they could do the same thing. They could do it a little bit faster and a little better on national TV, and 'I'm sorry, Ken. We don't need you anymore'.[31]

Jennings deflects from the feeling of obsolescence by reverting to ironic humour that revalues his own achievement as something that requires millions of dollars to emulate. All this may well be an attempt to give the whole situation the calming sense of normality and inconspicuousness that the oligarchic principle of technology and our own self-surrender to machines so effectively generate. In the postscript to *The Obsolescence of Human Beings 2* Anders reminds us why such 'gallows humour' should not prevent us from taking a journey into the inconspicuously exotic world of our digital lives. One ought to think of these words while 'doing nothing' when interacting with a machine:

> Our technological environment, the world of our devices, which consists of monsters, is either featureless, nondescript and inconspicuous, or it is deliberately kept out of sight. Anyone who has imagination to rely on instead of eyes, which nowadays no longer are of use, sees that it is precisely this invisibility, the invisibility of the monsters, which is the monstrosity of today. This invisibility transforms us into beings whose lives blindly pass by the machines, because we still have a most obsolete trust in our eyes. This at least, until the day when we will no longer have any time to realise that today our eyes cannot be trusted.[32]

In our ever more un-transparent times, or so I hope to have shown with this book, Anders's approach to thinking can contribute to a highly critical perspective, precisely because it invites us to resist, counter and bite back at the exaggerated images with which he presents us. It is true, that scrutiny of Anders's arguments at times may reveal that some of his images and propositions are, as Rudolf Burger puts it, 'only true as exaggerations'.[33] But they often hit a raw nerve. It is in this very impulse to resist Anders's pessimism that we are invited to contemplate the formative power of habits and are reminded that, for better or worse, 'what is does not have to be'.

I will give the final word to Anders, who in his defence of exaggeration as a method asks a question that will hopefully act as an invitation for further engagements with his highly original approach: 'Philosophers who refute exaggeration as unserious because they are used to work with their naked eyes – and most of them do so of course – are no less obsolete and ridiculous than a virologist would be if he refused microscopes in order to conduct "virology with the naked eye." Are viruses as big as they appear to the naked eye? If this were the case, they would not be at all. Are they not rather as big as they appear amplified through the microscope? Or are they not even bigger than this, because they are still incomparably more dangerous than their image discloses? If one were to screen the devastating workings of viruses a million times amplified in a film, would this amplification of the format also co-exaggerate the danger? Or would the danger here rather become visible for the first time?'[34]

NOTES

1. Anders, *Lieben gestern: Notizen zur Geschichte des Fühlens* (München: C.H. Beck, 1997), 13.
2. Anders, *Antiqueirtheit 2*, 423.
3. The Future of Life Institute, 'Autonomous Weapons: An Open Letter from AI & Robotics Researchers.' The letter can be accessed (and signed!) via http://futureoflife.org/open-letter-autonomous-weapons/.
4. For a discussion of drone warfare and violence that draws on Anders's 'On Promethean Shame' see Elke Schwarz, *The Biopolitical Condition: Re-Thinking the Ethics of Political Violence in Life-Politics* (PhD thesis 2013), especially chapter 7. http://etheses.lse.ac.uk/760/.
5. Erik Brynjolfsson and Andrew McAfee, *The Second Machine Age: Work, Progress and Prosperity in a Time of Brilliant Technologies* (New York: W.W. Norton & Company, 2014), 9.
6. David H. Autor, Frank Levy and Richard J. Murnane, 'Computer Based Technological Change and Skill Demands: Reconciling Perspectives of Economists and Sociologists', *Low-Wage America: How Employers are Reshaping Opportunity in the Workplace*, ed. Eileen Applebaum, Annette Bernhardt and Richard J. Murnane (New York: Russell Sage Foundation, 2004), 121–54, 122.
7. Turkle, *Alone Together*, 131. ELIZA is the name of a basic computer programme developed by Joseph Weizenbaum in the 1960s to study the interaction of humans and computers. 'The most convincing results came from a rule set,' as John Bohannon explains, 'that simulated a psychotherapist' by transforming statements into questions: 'In response to a patient saying, "I feel helpless," the computer might respond "Why do you feel that way?."' See John Bohannon, 'The Synthetic Therapist', *Science* 349.6245 (17 July 2015), 250–51. In order to keep the conversation going the human must actively adapt their way of speaking so as to not trip up the machine. The ELIZA effect names our willingness to behave in a way that helps us be tricked into believing that we are interacting with an empathetic entity that understands our concerns. Whereas Bohannon celebrates the potential of this effect, Turkle

is more sceptical: 'Over years and with some reluctance I came to understand that ELIZA's popularity revealed more than people's willingness to talk to machines; it revealed their reluctance to talk to other people. [...] The idea of an attentive machine provides the fantasy that we can escape each other. [...] Trust in ELIZA does not speak to what we think ELIZA will understand but to our lack of trust in the people who might understand' (*Alone Together*, 282).

8. See Anders, 'On Promethean Shame', (ch1) 82*.

9. Anders, 'On Promethean Shame', (ch1) 50*.

10. See Evgeny Morozov, *To Save Everything Click Here: Technology, Solutionism and the Urge to Fix Problems that don't Exist* (London: Penguin, 2013), esp. 171–2.

11. Roland Barthes, 'The Death of the Author', *Image, Music, Text*, trans. Stephen Heath (London: Fontana, 1977), 142–8, 148.

12. See Ivan Illich, 'Shadow – Work', *Philosophica* 26.2 (1980), 7–46.

13. Tiziana Terranova, 'Free Labor, Producing Culture for the Digital Economy', *Social Text* 18. 2 (2000), 33–57.

14. Anders, 'On Promethean Shame', (ch1) 78*.

15. IBM's Watson is the go-to example for a wide range of reflections about the future capabilities and uses of digital technology. My discussion responds to the description offered in *The Second Machine Age*, esp. 24–30.

16. Brynjolfsson and McAfee, *The Second Machine Age*, 64–5.

17. Anders, 'die Antiquiertheit der Privatheit', *Antiquiertheit* 2, 210–46. All references are from this edition and are given in the text. A translation of this important essay is forthcoming and will be published in *CounterText: A Journal for the Study of the Post-Literary* 3.1 (2017).

18. In the 1979 Preface to the 5th edition of *The Obsolescence of Human Beings 1*, Anders commented on the prophetic quality of his writings in a manner that seems highly apt in the context of the question of privacy: 'This volume, completed over a quarter century ago, does not seem outdated to me, but now more topical than then. This, however, says nothing for the quality of my analyses. It speaks, rather, emphatically against the quality of the situation of the world and humanity today.' See Anders, *Antiquiertheit 1*, vii.

19. Lanier, *Who Owns the Future?* 11.

20. Louis Althusser, 'Ideology and Ideological State Apparatuses (Notes towards an Investigation)', *Lenin and Philosophy and Other Essays*, trans. Ben Brewster (London: NLB, 1971), 121–73, 161.

21. Althusser, 'Ideology', 127.

22. Incidentally, big data is often presented as a tool to overcome the limited sight of human beings, because algorithms can establish patterns where humans cannot. While introducing their popular best-seller *Big Data*, Viktor Mayer-Schönberger and Kenneth Cukier even arrive at the perfect counter image to Anders's own definition of exaggeration and the suggestion that imagination needs to act as our telescope: 'Just as the telescope enabled us to comprehend the universe and the microscope allowed us to understand germs, the new techniques for collecting and analysing huge bodies of data will help us make sense of our world in ways we are just starting to appreciate.' See Mayer-Schönberger and Cukier, *Big Data: A Revolution that Will Transform How We Live, Work and Think* (London: John Murray, 2009), 7.

23. Anders, 'On Promethean Shame', (ch1) 42*.

24. Anders, *Antiquiertheit 2*, 145.
25. Turkle, *Alone Together*, 16.
26. Anders, *Antiquiertheit 2*, 55.
27. Anders, *Mensch ohne Welt*, xiii.
28. The endorsement is by Nicholas Negroponte, cofounder of the MIT Media Lab, founder of One Laptop per Child, and author of *Being Digital*, http://andrewmcafee.org/the-second-machine-age-the-book-and-the-blurbs/.
29. Lanier, *Who Owns the Future?*, xi.
30. Anders, *Antiquiertheit 1*, 311.
31. Ken Jennings, 'Watson, Jeopardy and Me, the Obsolete Know-It-All', accessed 9 March 2016, http://www.ted.com/talks/ken_jennings_watson_jeopardy_and_me_the_obsolete_know_it_all/transcript?language=en.
32. Anders, *Antiqueirtheit 2*, 423–4.
33. Rudolf Burger, 'Philosophie des Aufschubs', *Günther Anders kontrovers*, 252–66, 254.
34. Anders, *Philosophische Stenogramme*, 65.

Bibliography

Adorno, Theodor. "Free Time." In *The Culture Industry: Selected Essays on Mass Culture*, edited by J. M. Bernstein, 187–97. London: Routledge, 1991.

Adorno, Theodor, Günther Anders, Bertholt Brecht, Hans Eisler, Max Horkeimer, Herbert Markuse, and Ludwig Markuse. "Diskussionen aus einem Seminar über die Theorie der Bedürfnisse." In *Max Horkheimer. Gesammelte Schriften Band 12: Nachgelassene Schriften 1931–1949*, transcribed by Friedrich Pollock, Hans Reichenbach und Berthold Viertel, 559–86. Frankfurt am Main: Fischer, 1985.

Aeschylus. *Prometheus Bound.* Translated by A. J. Podlecki. Oxford: Aris & Phillips, 2005.

Althusser, Louis. "Ideology and Ideological State Apparatuses (Notes towards an Investigation)." In *Lenin and Philosophy and Other Essays*, translated by Ben Brewster, 121–73. London: NLB, 1971.

Anders, Günther. *Die Antiquiertheit des Menschen 1: Über die Seele im Zeitalter der zweiten industriellen Revolution* [The obsolescence of human beings 1: On the soul in the age of the second industrial revolution], 3rd ed. München: C.H. Beck, 2010.

———. *Die Antiquiertheit des Menschen 2: Über die Zerstörung des Lebens im Zeitalter der dritten industriellen Revolution* [The obsolescnece of human beings 2: On the destruction of life in the age of the third industrial revolution], 3rd ed. München: C.H. Beck, 2002.

———. *Die Atomare Drohung: Radikale Überlegungen zum atomaren Zeitalter* [The nuclear threat: radical reflections on the nuclear age], 7th ed. München: C.H. Beck, 2003.

———. "Being without Time: On Beckett's Play *Waiting for Godot.*" In *Samuel Beckett: A Collection of Critical Essays*, edited and translated by Martin Esslin, 140–51. Eaglewood Cliffs: Prentice-Hall, 1965.

———. *Besuch im Hades: Auschwitz und Breslau 1966. Nach "Holocaust" 1979* [Visit to Hades: Auschwitz and Breslau 1966. After "Holocaust" 1979], 3rd ed. München: C.H. Beck, 1996.

———. *Der Blick vom Mond: Reflexionen über Weltraumflüge* [The view from the moon: Reflections on spaceflights], 2nd ed. München: C.H. Beck, 1994.

———. *Burning Conscience: The Case of the Hiroshima Pilot, Claude Eatherly, Told in his Letters to Gunther Anders*. New York: Monthly Review Press, 1962.

———. "The Émigré." In *The Life and Work of Günther Anders*, edited by Günter Bischof, Jason Dawsey and Bernhard Fetz, translated by Otmar Binder, 171–86. Innsbruck: Studienverlag, 2014.

———. *Endzeit und Zeitende: Gedanken über die atomare Situation* [re-printed as *Die Atomare Drohung*]. München: C.H. Beck, 1972.

———. *Gewalt—ja oder nein. Eine notwendige Diskussion* [Violence yes or no? A necessary discussion]. Edited by Manfred Bissinger. München: Knauer, 1987.

———. *Hiroschima ist überall* [Hiroshima is everywhere]. München: C.H. Beck, 1982.

———. "Homeless Sculpture." *Philosophy and Phenomenological Research* 5, no. 2 (1944): 293–307.

———. *Kafka: Pro und Contra: Die Prozessunterlagen*. München: C.H. Beck, 1951.

———. *Kafka*. Translated by A. Steer and A. K. Thorlby. London: Bowes & Bowes, 1960.

———. *Ketzereien* [Heresies]. München: C.H. Beck, 1996.

———. *Die Kosmologische Humoreske und andere Erzählungen* [The cosmological humoresque and other stories]. Frankfurt am Main: Suhrkamp, 1978.

———. *Lieben gestern: Notizen zur Geschichte des Fühlens* [To be in love, yesterday: Notes toward the history of feeling], 3rd ed. München: C.H. Beck, 1997.

———. *Mensch ohne Welt: Schriften zur Kunst und Literatur* [Human without Wolrd: Writings on Art and Literature], 2nd ed. München: C.H. Beck, 1993.

———. "The Pathology of Freedom: An Essay on Non-Identification." Translated by Katharine Wolfe. *Deleuze Studies* 3, no. 2 (2009): 278–310.

———. *Philosophische Stenogramme* [Philosophy in shorthand], 3rd ed. München: C.H. Beck, 2002.

———. "The Pseudo-Concreteness of Heidegger's Philosophy." *Philosophy and Phenomenological Research* 8, no. 3 (March 1948): 337–71. Published in name: Günther Stern.

———. "Reflections on the H Bomb." Translated by Norbert Guterman. *Dissent* 3, no. 2 (Spring 1956): 146–55.

———. *Tagebücher und Gedichte* [Dairies and poems]. München: C.H. Beck, 1985.

———. *Tagesnotizen* [Daily notes]. Frankfurt am Main: Suhrkamp, 2006.

———. "Theses for the Atomic Age." Translated by Günther Anders. *Massachusetts Review* 3, no. 3 (Spring 1962): 493–505.

———. *Über Heidegger* [On Heidegger]. Edited by Gerhard Oberschlick. München: C.H. Beck, 2001.

———. "Une interprétation de l' a posteriori" [An interpretation of the a posteriori]. Translated by Emmanuel Levinas. *Recherches Philosophiques* 4 (1934): 65–80. Published in name: Günther Stern.

———. "Unverständlich" [Incomprehensible]. *Die Zeit* 18 (1988).

———. *Visit beutiful Vietnam: ABC der Aggressionen heute* [Visit beautiful Vietnam: today's alphabet of aggression]. Köln: Pahl-Rugenstein, 1968.

———. *Wir Eichmannsöhne* [We, the sons of Eichmann], 3rd ed. München: C.H. Beck, 2002.

———. "The World as Phantom and Matrix." Translated by Norbert Guterman. *Dissent* 3, no. 1 (Winter 1956): 14–24.

Arendt, Hannah and Günther Stern (Anders). "Rilkes *Duineser Elegien*" [Rilke's *Duino Elegies*]. *Neue Schweizer Rundschau* 11 (1930): 855–71.

Artaud, Antonin. *The Theatre and its Double*. Translated by Victor Corti. Richmond: Oneworld Classics, 2010.

Autor, David H., Frank Levy, and Richard J. Murnane. "Computer Based Technological Change and Skill Demands: Reconciling Perspectives of Economists and Sociologists." In *Low-Wage America: How Employers are Reshaping Opportunity in the Workplace*, edited by Eileen Applebaum, Annette Bernhardt and Richard J. Murnane, 121–54. New York: Russell Sage Foundation, 2004.

Babich, Babette. "*O, Superman!* Or Being Toward Transhumanism: Martin Heidegger, Günther Anders, and Media Aesthetics." *Divinatio* 36 (2013): 42–99.

Bardini, Thierry. *Junkware*. Minneapolis: University of Minnesota Press, 2011.

Barthes, Roland. "The Death of the Author." In *Image, Music, Text*, translated by Stephen Heath, 142–48. London: Fontana, 1977.

Bauman, Zygumnt. *Consuming Life*. Cambridge: Polity, 2007.

Benjamin, Walter. "Poverty and Experience." In *Selected Writings Volume 2: 1927–1934*, edited by Michael W. Jennings, Howard Eiland and Gary Smith, translated by Rodney Livingstone, 731–35. Cambridge, MA: Harvard University Press, 1999.

Bischof, Günther, Jason Dawsey, and Bernhard Fetz, eds. *The Life and Work of Günther Anders*: Émigré, Iconoclast, Philosopher, Man of Letters. Innsbruck: Studienverlag, 2014.

Bohannon, John. "The Synthetic Therapist." *Science* 349, no. 6245 (17 July 2015): 250–51.

Bradley, Arthur. *Originary Technicity: The Theory of Technology from Marx to Derrida*. Basingstoke: Palgrave MacMillan, 2011.

Brassier, Ray. "Prometheansim and its Critics." In *The Accelerationist Reader*, edited by Robin Mackay and Armen Avanessian, 467–87. Berlin: Merve, 2014.

Bryden, Ronald. "The Myth of Major Eatherly." *The Spectator*. February 16, 1962, 5.

Brynjolfsson, Erik and Andre McAffee. "The Great Decoupling." *International Herald Tribune*. December 12, 2012, 8.

———. *The Second Machine Age: Work, Progress and Prosperity in a Time of Brilliant Technologies*. New York: W.W. Norton & Company, 2014.

Buck-Morss, Susan. "Aesthetics and Anaesthetics: Walter Benjamin's Artwork Essay Reconsidered." *October* 62 (1992): 3–41.

Burger, Rudolf. "Philosophie des Aufschubs" [Philosophy of postponment]. In Liessmann, *Günther Anders kontrovers*, 252–66.

Crogan, Patrick and Samuel Kinsley. "Paying Attention: Towards a Critique of the Attention Economy." *Culture Machine* 13 (2012): 1–29.

Deleuze, Gilles. "Postscript on the Societies of Control." In *Gilles Deleuze: Negotiations, 1972-1990*, translated by Martin Joughin, 177–82. New York: Columbia University Press, 1995.

van Dijk, Paul. *Anthropology in the Age of Technology: The Philosophical Contribution of Günther Anders*. Translated by Frans Kooymans. Amsterdam: Rodopoi, 2000.

Dougherty, Carol. *Prometheus*. New York: Routledge, 2006.

Dries, Christian. *Die Welt als Vernichtungslager. Eine kritische Theorie der Moderne im Anschluss an Günther Anders, Hannah Arendt und Hans Jonas* [The world as a termination camp. A critical theory of modernity following Günther Anders, Hannah Arendt and Hans Jonas]. Bielefeld: transcript, 2012.

Dupuy, Jean-Pierre. "Some Pitfalls in the Philosophical Foundations of Nanoethics." *Journal of Medicine and Philosophy* 32 (2007): 237–61.

———. *The Mark of the Sacred*. Translated by M. B. Debevoise. Stanford: Stanford University Press, 2013.

Emerson, Ralph Waldo. "Ode Inscribed to William S. Channing." In *Emerson's Complete Works, Volume 9: Poetry*, edited by James Elliot Cabot, 71–74. London: Waverley, 1893.

Esposito, Robert. *Immunitas*. Translated by Zakiya Hanafi. Cambridge: Polity, 2013.

Freud, Sigmund. "The Uncanny." In *The Standard Edition of the Complete Psychological Works of Sigmund Freud Volume XVII: An Infantile Neurosis and Other Works*, edited by James Strachey, in collaboration with Anna Freud, translated by Alix Strachey, assisted by Alix Strachey and Alan Tyson, 219–52. London: Hogarth, 1955.

Fuld, Werner. "Zwischen Film und Bombe. Die Kontinuität des Andersschen Denkens" [Between film and bomb. The continuity of Anders's thought]. In Liessmann, *Günther Anders kontrovers*, 114–23.

The Future of Life Institute. "Autonomous Weapons: An Open Letter from AI & Robotics Researchers," accessed February 10, 2016, http://futureoflife.org/open-letter-autonomous-weapons.

Gibson, William. "Google's Earth." *New York Times*. 1st of September, 2010, A23.

Hadot, Pierre. *The Veil of Isis: An Essay on the History of the Idea of Nature*. Translated by Michael Chase. Cambridge, MA: Harvard University Press, 2006.

Hans-Martin Lohmann. "'Glückliche Zeiten als es noch Kriege gab.' Variationen zu einem Satz von Günther Anders" ['Happy times as wars still existed': Variations on a sentence by Günther Anders]. In Liessmann, *Günther Anders kontrovers*, 211–20.

Hauskeller, Michael. *Sex and the Posthuman Condition*. Basingstoke: Palgrave MacMillan, 2014.

Hayles, Catherine. *How We Think: Digital Media and Contemporary Technogenesis*. Chicago: University of Chicago Press, 2012.

Heidegger, Martin. *Sein und Zeit*. Tübingen: Max Niemeyer, 2006.

———. *Unterwegs zur Sprache*. Stuttgart: Klett-Cotta, 2007.

Hesiod. *Theogony and Works and Days*. Translated by M. L. West. Oxford: Oxford University Press, 1999.

Hesiod. *Theogony and Works and Days*. Translated by Catherine M. Schlegel and Henry Weinfield. Ann Arbor: University of Michigan Press, 2006.

Illich, Ivan. "Shadow—Work." *Philosophica* 26, no. 2 (1980): 7–46.

Jaspers, Karl. *Philosophy Vol. 2: The Illumination of Existence*. Translated by E. B. Ashton. Chicago: Chicago University Press, 1969.

Jennings, Ken. "Watson, Jeopardy and Me, the Obsolete Know-It All," http://www.ted.com/talks/ken_jennings_watson_jeopardy_and_me_the_obsolete_know_it_all/transcript?language=en.

Jungk, Robert. *Die Zukunft hat schon begonnen* [The future has already begun]. Stuttgart: Scherz und Goverts, 1952.

Kant, Immanuel. "Anthropology from a Pragmatic Point of View." In *The Cambridge Edition of the Works of Kant: Anthropology, History, and Education*, edited by Günther Zöller and Robert Louden, translated by Robert Louden, 227–429. Cambridge: Cambridge University Press, 2007.

Kracauer, Siegfried. *Die Angestellten*. Frankfurt am Main: Suhrkamp, 1971.

———. "Photography." *The Mass Ornament: Weimar Essays*. Edited and translated by Thomas Levin, 47–64. Cambridge, MA: Harvard University Press, 1995.

———. *The Salaried Masses: Duty and Distraction in Weimar Germany*. Translated by Quintin Hoare. London: Verso, 1998.

Lanier, Jaron. *You are Not a Gadget*. London: Penguin, 2007.

———. *Who Owns the Future?* London: Penguin, 2013.

Leroi-Gourhan, André. *Gesture and Speech*. Translated by Anna Bostock Berger. Cambridge, MA: MIT Press, 1993.

Liessmann, Konrad Paul. *Günther Anders zur Einführung* [An introduction to Günther Anders]. Hamburg: Junius, 1988.

———, ed. *Günther Anders kontrovers* [Günther Anders controversial]. München: C.H. Beck, 1992.

———. *Günther Anders: Philosophieren im Zeitalter der technologischen Revolution* [Günther Anders: philosophy in the age of the technological revolution]. München: C.H. Beck, 2002.

———. "Thought after Auschwitz and Hiroshima: Günther Anders and Hannah Arendt." Translated by Andrew Smith. *Enrahonar* 46 (2011): 123–35.

———. "Die Gerätewelt. Günther Anders und die Phänomenologie der Dinge" [The machine-world. Günther Anders and the phenomenology of things]. In *Kraft der Dinge: Phänomenologische Skizzen*, edited by Iris Därman, assisted by Rebekka Ladewig, 69–82. Paderborn: Wilhelm Fink, 2014.

Lyotard, Jean-François. "Note on the Meaning of 'Post-'." In *Postmodernism*, edited by Thomas Docherty, 47–50. Hemel Hempstead: Harvester Wheatsheaf, 1993.

Malabou, Catherine. *What Should We Do with Our Brain?* Translated by Sebastian Rand. New York: Fordham University Press, 2008.

Mayer-Schönberger, Viktor and Kenneth Cukier. *Big Data: A Revolution That Will Transform How We Live, Work and Think*. London: John Murray, 2013.

Morozov, Evgeny. *To Save Everything Click Here: Technology, Solutionism and the Urge to Fix Problems that Don't Exist*. London: Penguin, 2014.

Müller, Christopher. "Desert Ethics: Technology and the Question of Evil in Günther Anders and Jacques Derrida." *Parallax* 21, no. 1 (2015): 42–57.

———. "We are Born Obsolete: Günther Anders's (Post)humanism," http://critical-posthumanism.net/?page_id=433.

Müller, Marcel. *Von der Weltfrendheit zur Antiquiertheit: Philosophische Anthropologie bei Günther Anders* [From estrangment to the world to obsolescence: Günther Anders's philosophical anthropology]. Marburg: Tectum, 2012.

Nancy, Jean-Luc. "A Finite Thinking." *A Finite Thinking*. Edited by Simon Sparks, translated by Edward Bullard, Jonathan Derbyshire and Simon Sparks, 3–30. Stanford: Stanford University Press, 2003.

Nancy, Jean-Luc. *After Fukushima: The Equivalence of Catastrophes*. Translated by Charlotte Mandell. New York: Fordham University Press, 2015.

Nancy, Jean-Luc and Aurélien Barrau. *What's Theses Worlds Coming To?* Translated by Travis Holloway and Flor Méchain. New York: Fordham University Press, 2015.

Nietzsche, Friedrich. *Thus Spoke Zarathustra: A Book for All and None*. Translated by Adrian del Caro. Cambridge: Cambridge University Press, 2006.

Plato. "Protagoras." In *Gorgias, Menexenus, Protagoras*, edited by Malcolm Schonfield, translated by Tom Griffith, 143–204. Cambridge: Cambridge University Press, 2010.

Reimann, Werner. "Nihilismus und Scham" [Nihilism and shame]. In Liessmann, *Günther Anders Kontrovers*, 57–71.

———. *Verweigerte Versöhnung: Zur Philosophie von Günther Anders* [Reconciliation denied: on the philosophy of Günther Anders]. Wien: Passagen, 1990.

Scheler, Max. "Über Scham und Schamgefühl" [On shame and feeling shame]. In *Schriften aus dem Nachlass Band 1: Zur Ethik und Erkenntnislehre*, edited by Manfred S. Frings, 65–154. Bern: Francke, 1957.

Schelling, F. W. J. *Philosophical Investigations into the Nature of Human Freedom*. Translated by Jeff Love and Johannes Schmidt. Albany: State University of New York Press, 2006.

Schopenhauer, Arthur. *Parerga und Paralipomena II: Zweiter Teilband*. Zürich: Diogenes, 1977.

Schraube, Ernst. "'Torturing Things Until They Confess': Günther Anders's Critique of Technology." *Science as Culture* 14, no. 1 (2005): 77–85.

Schröder, Klaus Albrecht. "Die Genese von Günther Anders' Medienkritik" [The origins of Günther Anders's media-critique]. In Liessmann, *Günther Anders Kontrovers*, 124–35.

Schubert, Elke. *Günther Anders*. Hamburg: Rowohlt, 1992.

Schwarz, Elke. *The Biopolitical Condition: Re-Thinking the Ethics of Political Violence in Life-Politics*. PhD thesis, The London School of Economics and Political Science (LSE), 2013.

Stiegler, Bernhard, ed. *Günther Anders Antwortet: Interviews und Erklärungen* [Günther Anders responds: Interviews and explanations]. Berlin: Tiamat, 1987.

———. *Technics and Time Volume 1: The Fault of Epimetheus*. Translated by Richard Beardsworth and George Collins. Stanford: Stanford University Press, 1998.

———. "Technics of Decision: An Interview." Translated by Sean Gaston. *Angelaki* 8, no. 2 (2003): 151–68.

———. "Anamnēsis and Hypomnēsis: The Memories of Deisre." In *Technicity*, edited by Louis Armand and Arthur Bradley, translated by François-Xavier Gleyzon, 15–41. Prague: Litteraria Pragensia, 2006.

———. *The Decadence of Industrial Democracies: Disbelief and Discredit, Volume 1*. Translated by Daniel Ross and Suzanne Arnold. Cambridge: Polity, 2011.

———. *What Makes Life Worth Living: On Pharmacology*. Translated by Daniel Ross. Cambridge: Polity, 2013.

Terranova, Tiziana. "Free Labour, Producing Culture for the Digital Economy." *Social Text* 18, no. 2 (2000): 33–57.

Truman, Harry. "Statement by the President of the United States," http://www.trumanlibrary.org/whistlestop/study_collections/bomb/large/documents/index.php.

Turkle, Sherry. *Alone Together: Why We Expect More from Technology and Less from Each Other*. New York: Basic, 2011.

Waugh, Evelyn. *The Loved One: An Anglo-American Tragedy*. London: Penguin, 2000.

Williams, Alex and Nick Srnicek. "#Accelerate: Manifesto for an Accelerationist Politics." In *The Accelerationist Reader*, edited by Robin Mackay and Armen Avanessian, 347–62. Berlin: Merve, 2014.

Wills, David. *Dorsality: Thinking Back through Technology and Politics*. Minnesota: Minnesota University Press, 2008.

Wittulski, Eckhard. "Der tanzende Phänomenologe" [The dancing phenomenologist]. In Liessmann, *Günther Anders kontrovers*, 17–33.

Woessner, Martin. *Heidegger in America*. Cambridge: Cambridge University Press, 2011.

Wolin, Richard. *Heidegger's Children*. Princeton: Princeton University Press, 2001.

Index

About the Author

Christopher John Müller is Honorary Research Associate of the School of English, Communication and Philosophy at Cardiff University and Associate Teacher at the University of Bristol. He completed his doctoral thesis in 2013 at the Centre for Critical and Cultural Theory, Cardiff University. His work draws on Literature, Philosophy, Cultural Studies and Critical Theory to address the manner in which technological and linguistic structures shape human perception, agency and interaction. His recent publications include 'Desert Ethics: Technology and the Question of Evil in Günther Anders and Jacques Derrida', *Parallax* 21, no. 1 (2015): 42–57 and 'Style and Arrogance: The Ethics of Heidegger's Style', *Style in Theory: Between Literature and Philosophy*, ed. Ivan Callus, Gloria Lauri-Lucente and James Corby (2013), 141–62. Further translations of works by Günther Anders are in preparation. More information can be found on his personal website www.cjmuller.co.uk.

www.ingramcontent.com/pod-product-compliance
Lightning Source LLC
LaVergne TN
LVHW101448200926
848483LV00001B/89